GERMAN VEHICLES IN WORLD WAR II
Cars of the Wehrmacht
A Photo Chronicle

Reinhard Frank

The first light of morning near Blonie, west of Warsaw, on September 15, 1939. The vehicles of the 14th (Panzerjäger) Company of I.R. 82 (31st I.D.) stand close together, arranged in columns, in fully coverless country, an inviting target for bombs and shells of all sizes and shapes!

The unit is equipped with "medium gl. passenger vehicle with towing apparatus (Kfz. 12) with chassis of the medium Pkw (o)" to tow its 3.7 cm antitank guns. As usual in the Wehrmacht, these were mainly the products of a wide variety of manufacturers: at left the comparatively uncommon Phänomen Granit Type 25 H with open box body, in the center foreground a Mercedes-Benz Type Stuttgart, at right front a Wanderer W 11, and in the center several Adler Type 3 Gd.

GERMAN VEHICLES IN WORLD WAR II
Cars of the Wehrmacht
A Photo Chronicle

Reinhard Frank

Schiffer Military History
Atglen, PA

Minsk, summer 1944; the Russians have broken through! The encirclement and complete destruction of Army Group Center is underway. Back-line services, officials and Russian citizens leave the threatened city in flight. Good for those who have seized a driveable vehicle. – Volkswagen, DKW, GAZ, Ford, gasoline or wood-gas powered – anything will do, as long as runs and takes you far enough west!
The VW Type 82 E and the DKW Reichsklasse (Meisterklasse) belonged to the Reich Transit Authority in Minsk; the Ford V 300 S truck with the puzzling registration MH-5334 bears the lettering NSKK 4/38 (Sturm 4/Motorstandarte 38?) and an oak leaf, such as group leaders wore on their collar patches.

Translated from the German by Dr. Edward Force.

This book was originally published under the title,
Personenkraftwagen der Wehrmacht,
by Podzun-Pallas Verlag.

Copyright © 1994 by Schiffer Publishing Ltd.
Library of Congress Catalog Number: 94-66975

All rights reserved. No part of this work may be reproduced or used in any forms or by any means – graphic, electronic or mechanical, including photocopying or information storage and retrieval systems – without written permission from the copyright holder.

Printed in the United States of America.
ISBN: 0-88740-687-4

We are interested in hearing from authors with book ideas on related topics.

Published by Schiffer Publishing Ltd.
77 Lower Valley Road
Atglen, PA 19310
Please write for a free catalog.
This book may be purchased from the publisher.
Please include $2.95 postage.
Try your bookstore first.

Contents

Foreword	7
Cars of the Wehrmacht	9
Historical Development	9
From 1885 to 1918	9
From 1919 to 1933	12
From 1933 to 1945	15
Passenger Vehicles in Action	19
Organization and Classification	20
Load Classes	22
German Cars	28
Uniform Vehicles	28
The Light Uniform Pkw	28
The Medium Uniform Pkw	37
The Heavy Uniform Pkw	51
The Kübelwagen	61
Three-Axle Cars	92
Requisitioned Cars	96
Captured Cars	151
France	152
Britain	170
Russia	174
U.S.A	175
Appendix	194
The Towing Axle	194
Road Signs	195

One car after another on the square by the railroad station in Kursk. A variety in field gray, that makes the heart of the auto enthusiast beat faster today, but drove every quartermaster to the brink of despair then ... (Autumn 1942, Field Police Unit 521 (mot.))

Foreword

After the book on the trucks of the Wehrmacht, here is the book on its passenger vehicles.

From the first to the last day of the war, these vehicles had their jobs to do on all fronts and on the home front – no matter whether or not they were at all suited to service in sand, slime or snow. As the pictures show, along with the "military" Kübelwagen, an unbelievable variety of civilian German and foreign passenger cars went with the troops – an undeniable indication of the notorious lack of vehicles in the Wehrmacht. With few exceptions, the off-road capability of these vehicles was modest in the extreme, and for the most part it did not exist at all. The snug interiors and modest controls were in no way built with a soldier in mind, especially one bundled up for the winter and wearing heavy boots. Very few cars of those times possessed interior heating, so that there were often cases of severe frostbite (especially to the driver's legs). In the summer, dust and sand soon infiltrated the unprotected motors, bearings and joints and turned them to scrap. Chassis, springs and axles generally could not stand the bad roads for long and broke down – the field repair shops never had reason to complain of a lack of work! The lack of any armoring necessarily resulted in frequent damage in the front areas, as well as in the hinterlands, thanks to partisans and low-flying enemy planes.

Most of the photos were provided by former soldiers. Shortcomings in terms of sharpness, lighting and choice of subjects must be accepted at times if one is not to do without irreplaceable pictures.

The identification of the vehicles was accomplished by consulting all available books on makes and models. As every experienced car spotter knows, a precise identification of the model is often very difficult, since there were vast variations in the form of small series, special productions, variants, modifications, rebuilding and the like. Further information and constructive criticism from our readers are always welcome.

Special thanks go to my wife Gertrud and Mr. Henry Hoppe, who provided great assistance with his photo archives, and retired Oberst Gerhard Elser, who devoted a great deal of tedious work to gathering and evaluating important source material.

This is a technical book without any political standpoint. It concerns a time when millions of people lost their lives, their health, their homeland – let us remember that at this time.

Gilching, Autumn 1993

Reinhard Frank

German Abbreviations

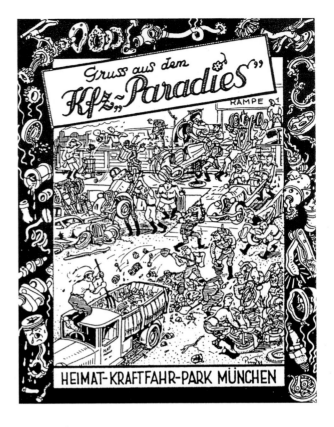

gl.	= off-road capable (geländegängig)
I.D.(mot.)	= Infantry Division (motorized)
I.R.	= Infantry Regiment
Kfz	= motor vehicle (Kraftfahrzeug)
Kw	= motor vehicle (Kraftwagen)
Lkw	= truck (Lastkraftwagen)
l.	= light (leicht)
m.	= medium (mittel)
(o)	= stock (handelsüblich)
Pkw	= car (Personenkraftwagen)
Sd.Anh.	= special trailer (Sonderanhänger)
Sd.Kfz.	= special vehicle (Sonderkraftfahrzeug)
s.	= heavy (schwer)
3-Tonner	= vehicle with 3-ton load or towing capability
(4 x 2)	= two of four wheels are driven

Note: Most abbreviations were used both with and without a period (Pkw. or Pkw). Naming of individual components was done extensively according to D 659/10 of April 15, 1943.

Photos and documents have been made available by

Ludwig Baer, Horst Beiersdorf, Friedrich Birkmeyer, Theodor Brandherm, Siegfried Bunke, Johann Dermleitner, Siegfried Ehrt, Gerhard Elser, Robert Emmert, Wolfgang Essler, Michael Foedrowitz, Hermann Freter, Heinz Friedrich, Sepp Herz, Helmut Heinrich, Henry Hoppe, Hansludwig Huber, Gerhard Kesenhagen, Karl-Heinz Köppl, Randolf Kugler, Michael Mair, Friedrich Masch, Ernst Obermaier, Peter Petrick, Horst Scheibert, Halwart Schrader, Franz Sindler, Johann Strauss, Helmuth Spaeter, Peter Taghorn, Hans Thudt, Bart Vanderveen, Francois Vauvillier, Heinz Wrobel, Albert Wörnle, Georg Wurm and many others.

Cars of the Wehrmacht

Historical Development

From 1885 to 1918

Over a hundred years have now passed since Karl Benz built the first usable motor car in 1885 and thus laid the cornerstone for the general motorization of Germany – with all its advantages and disadvantages. Naturally, the military soon showed a certain interest in the use of motor vehicles in army service, but the prevailing conservative outlook of the leadership allowed only a very hesitant introduction of this vehicle, which was still very unreliable and scarcely suitable for display in parades.

The first motor vehicle that saw experimental service with the Imperial Army was a rented Daimler truck in 1898. During Imperial maneuvers in 1899, along with various trucks, a passenger car was used successfully for the first time to transport higher officers and deliver messages.

The use of passenger cars was well known, of course, but such vehicles were obtained by the Army administration only in very small numbers (generally for high-level staff use). The truck was ranked as being of essential importance in wartime, and thus held absolute priority. In case of war, the need for "large passenger cars" and small two-seat "despatch cars" should be met, according to a section of the War-Waging Law of June 13, 1873, almost exclusively by requisitioning private vehicles.

In bad road conditions, most cars soon lost their momentum and had to wait patiently for a team of draft horses to tow them away – such as this landaulet from the fortress of Brest in 1915.

As for the use of the car, the Field Service Orders of March 22, 1908 state, among other things:

Passenger cars are issued to the higher staffs. In addition to transporting the staff, they serve to deliver important messages and orders. An officer is to be charged with the purposeful use of the cars.

On account of their great speed, the passenger cars form an exemplary means of carrying orders and information, especially in cases where personal understanding is important. But they are to be used only on good roads and only in areas secured by one's own troops. Their speed can only be utilized on the open road. They break down easily where careful maintenance is lacking.

When the war began in 1914, about 800 passenger cars went into the field with the motorized units (staff and cavalry vehicle columns, staff car pools, etc.). In the autumn of 1915, the troops were already using some 8600 passenger ve-

A right-hand-drive Protos car.

hicles – all of them requisitioned from civilian owners. The variety of these cars was naturally vast: the special development of a uniform army passenger vehicle did not take place, since the comparatively small automobile factories were fully occupied with their own production programs, and also because there was no central administration of production.

By the end of the war in 1918, the vehicles of the German army numbered:

12,000 passenger vehicles, 3200 ambulances, 25,000 trucks and 5400 motorcycles.

The Allied sea blockade had also resulted in a serious shortage of raw materials for the production of motor vehicles, plus a tremendous shortage of fuel, which could never be made up for by numerous, notorious substitute materials.

There were already captured cars in World War I: "The car of captured Englishmen in Finis during the spring offensive, March 22, 1918."

As this picture shows, the advantages of motorized, well-armed infantry (rather like the later armored grenadiers) were already recognized. The car is equipped with two comparatively rare 7.92 mm Bergmann 1915 machine guns. The picture could come from either the last weeks of World War I or the days of the free corps.

From 1919 to 1933

To end the postwar chaos within the country and at its borders, the government passed a "law to form a Reichswehr for the time being" on March 6, 1919, and a "Military Law" based on the Reich Constitution on March 23, 1921.

The use of passenger vehicles by the Reichswehr, was defined as follows by the Chief of Army Command, General Hans von Seeckt, in the order "Command and Combat of Fixed Weapons of 9/1/1921 (Appendix, Motor Vehicles, of 6/20/1923).

Passenger vehicles are used by the Army as: Small vehicles for reconnaissance and column use with open two- to six-seat bodies and 35 kph average and 60 to 70 kph top speeds.

Passenger vehicles for the higher staffs with open or closed six-seat body and 35 kph average and 70 to 100 kph top speeds.

Passenger vehicles make the upper and middle staffs mobile and allow them frequent, direct exchanges of ideas and a feeling of closeness to the troops.

The organization and equipping of the Reichsheer could not be done at any time according to the practical experience gained in the war, but always only according to the restrictions of the Treaty of Versailles and the meager budget of the Reich.

The meager strength of the Reichsheer must be at least partially compensated for by increasing its mobility. For that reason, making the troops mobile and enhancing their mobility by means of motor vehicles are especially important.

Thus it was stated very purposefully in HDv 472 of August 29, 1924, under section 470. For lack of money, it was only advised at first to put "militarily useful" bodies on normal production chassis. Only vehicles with rear-wheel drive and (with few exceptions) rigid axles were involved.

To confuse the Allied control organizations, the Reichswehr drilled with a variety of unobtrusive private cars, such as this Pluto (made under license from Amilcar, 4 cyl., 1.1 liter, 20 HP) built from 1924 to 1927, with license plates for the province of East Prussia.

As of 1929, so called "Off-Road Passenger Vehicles" were introduced: standard car chassis (4x2) with a special body that was fully open on the sides for quick action and carried a folding canvas top for (insufficient) protection from the weather. Increased ground clearance was attained by using larger wheels, stronger springs, and moving or protecting various components such as the exhaust pipe, muffler and brake system. In some cases, a locking differential and a rear axle with different ratios were installed. An obvious identifying mark were the seats in "box form" (a development of the Trutz firm of Coburg), which were supposed to give the passengers lateral support.

The designation of "Kübelsitzwagen" was soon shortened to "Kübelwagen" and then to the informal "Kübel."

Typical "Kübel cars of that time were the Mercedes-Benz "Stuttgart 260", the Adler "Favorit" and the Hanomag 4/20.

The development of a passenger vehicle with full off-road capability, with three axles and six seats, which was carried out by Selve, Horch and Daimler-Benz in 1925-26, brought only little success and was called off after a few test models were built.

The design had proved to be too complicated and expensive. Such vehicles were developed later on the basis of light truck chassis (such as the Krupp Protze).

Members of the Reichswehr in one of the rare three-axle cars, a "heavy off-road Pkw" made by Horch in 1926.

In 1930 the Army, according to H. A. Koch (Feldgrau, December 1964), had 406 cars and 1176 trucks. In particular:

14 cars and 98 trucks for 7 motorized batteries
21 cars and 273 trucks for 21 ambulance platoons
371 cars and 805 trucks for 7 motorized units (KfA)

It is very likely that there were also a number of "illicit" supplies of motor vehicles, which can no longer be determined.

The variety of motor vehicle types of all sorts in the Reichswehr was very great, since on account of the serious financial situation, contracts were given to as many manufacturers as possible. The use of standard production chassis was also very problematic, since every model change of a manufacturer led to a further addition to the list of types.

Even though the dummy tanks of the Reichswehr inspired many jokes, they were vitally important in training in terms of the command and tactical use of fast-moving units. This picture shows imitation armored vehicles on the chassis of the Adler Standard 6 car, with one vehicle already equipped with the radio apparatus and frame antenna of a command car. At right is an open Adler with the typical bucket seats. The variety of vehicles in the background is striking.

From 1933 to 1945

With the "seizure of power" in 1933 and the efforts to build up a modern military, a strong advance in the use of motor vehicles began. The long-sought goal of developing the army's own standardized passenger vehicle was now taken up at top speed.

The Army Weapons Office therefore made plans for three so-called "uniform chassis" for one light, one medium and one heavy passenger vehicle. Establishing norms and types constituted an effort to simplify production and make spare-parts supply easier. The chassis featured all-wheel drive and could be fitted with various bodies suited to their type of use.

The first uniform passenger vehicles reached the troops in 1937.

Unfortunately, they did not fulfill all the hopes that had been placed on them: they proved to be very heavy, need much maintenance, and be very expensive. Because of the lack of capacity in the overburdened automobile industry, no uniform motors could be developed, so that standard powerplants made by a variety of manufacturers, and not interchangeable, had to be relied on. Since there were also great numbers of variations within production series (which will be dealt with more fully in the chapter on the uniform passenger vehicle), the intended standardization was out of the question.

Only in June of 1940 did the first truly "uniform" passenger vehicle reach the troops: the "VW Kübel" of Professor Porsche. his vehicle, to be sure, had only rear-wheel drive, but thanks to its light weight, its sturdiness and reliability, it was far superior to the uniform passenger vehicles.

All of these efforts to standardize vehicle production were made on the initiative of the general command for motor vehicles (Chief of Official Group K in the OKH), Oberst/General Adolf von Schell.

He had the task of limiting the uneconomic variety of motor vehicles of all kinds (including aggregates and equipment) to its simplest terms in accordance with the "Schell Plan" which he established.

With the "Directive on the Simplification of the Motor Vehicle Industry" of March 2, 1939, for example, the number of civilian car types was supposed to be limited to only 31 acceptable basic models.

In the course of the war, according to the report of the GBK of November 1, 1941, the production of only one type of passenger vehicle (the Volkswagen) was planned in Germany, and three types in the protectorate.

It was stated in this report:

"All special designs of the Wehrmacht in the realm of wheeled motor vehicles were weeded out (some of them are still being concluded), and specifically, special designs for the light uniform passenger vehicle were replaced by the VW (two- and four-wheel drive). Special designs for the medium and heavy uniform passenger vehicle were replaced by vehicles on the chassis of the 1.5-ton truck (S or A type). The corresponding motor vehicles (VW, 1.5-ton, 3-ton and 4.5-ton) are built with two-wheel drive for the market, four-wheel drive for the Wehrmacht."

But despite all efforts, only a very few units could be equipped with a uniform supply of motor vehicles.

The German automobile industry, with its variety of independent (and independent-thinking) companies, had simply reached the limits of its capability.

Until the end of the war, the Wehrmacht was compelled to operate with an incredible variety of vehicles of German and foreign manufacture.

At the very beginning of the war, the Wehrmacht, according to Fritz Wiener, went into the field with 63% "completion vehicles" and only 37% special Wehrmacht vehicles.

"Completion" meant, in Wehrmacht language, bringing peacetime troop units up to wartime strength with men, animals, weapons and equipment, and setting up troop units not included in wartime structure (see Mueller-Hillebrand, "Das Heer 1933-45").

"Completion vehicles" were simply all those vehicles that had been confiscated from private owners, firms and agencies. The legal basis for taking them was the "law on procurement for defensive purposes of 7/13/1938 (Wehrleistungsgesetz)." As in similar authorizations made before World War I, the WLG established a "general procurement duty" for the purpose of the defense of the Reich." On the basis of this law, the OKH had already announced, on August 13, 1938, a Motor Vehicle Completion Directive HDv 167 within the general clarifications and military operational determinations of HDv g 167.

After the authorizations that were announced at various times, there followed, in the army instruction publications, clarifications on motor vehicle procurement on the basis of factory and rental contracts or for procurement to meet daily needs, about responsibilities, about the procurement of motor vehicle drivers, and the like.

On September 1, 1938, the "law on procurement for Reich duties" (Reichsleistungsgesetz) was finally proclaimed.

Paragraph 15 of the RLG stated, among other things:

The possessors of land, air and water vehicles of all kinds are obligated to make these available for use at the place where they are needed.

As places where land vehicles "were needed by the Wehrmacht", the WLG and RLG stipulated the defense replacement inspections (WEI) named by the defensive district commands.

According to HDv 75 NfD "Determinations for the maintenance of the Army in a wartime condition" of September 15, 1939, it was the duty of the WEI:
— to obtain all the motor vehicles in the home area, other than the Wehrmacht's own vehicles, according to HDv 157, namely to take possession of them officially, watch over them and disperse them for the purpose of general waging of the war;
— to identify vehicles still available after the establishment of the wartime basis of the Wehrmacht;
— to confiscate motor vehicles available from automobile dealers and agencies of manufacturers, and
— to report the complete results to the commander of the replacement army.

When the war began (or even shortly before), wartime economy measures against private transit took effect:
8/28/1939: Permit required for liquid fuels.
9/1/1939: Gasoline rationing, tank license cards, obligatory blacking-out of motor vehicles.

9/3/1939: All manufacturers were banned from selling and delivering motor vehicles. The Wehrmacht confiscated four-seat open cars.
9/6/1939: "Directive on the further use of motor vehicles": only vehicles that served "the public interest" were allowed. This was indicated by the small, very desired "red angle" on license plates, which 23% of all civilian vehicles received at first, later "considerably fewer."
9/11/1939: Confiscation of automobile tires, tire rationing for permitted vehicles.

It can be concluded that, according to the regulations, the Army acquired certain models as "Wehrmacht vehicles" and deliberately modified them. These included, for example, the Opel Kadett and Olympia (D 632/20 and 20b), Super 6 and Kapitän (D 632/49 a), and Ford V8 and Horch 830.

Rebuilding them resulted in either open Kübel-seat cars or closed vehicles with box bodies for use as command, communication or towing vehicles. Both types were equipped with shielded lights, storage cases, carbine and trenching-tool racks, and painted with whatever identifying marks were required. In addition to their civilian registration numbers, they bore:
– a white circle on the front and rear with the black letters "WH", according to the "Mobilization Plan for the Army" of 1937;
– white letters "WH", according to Army Directive B 1940, No. 95, of January 31, 1940 (in part also for action in Austria in 1930).

Only as ordered on July 3 and November 27, 1941 (Army Directive C 1941, No. 621 and 1060), were "completion vehicles" also given a white license plate.

On the basis of available documents, the following information about production and availability statistics can be given:
– Available passenger cars in the German Reich in 1938 (Source: "Schlag nach, 1939"):

Passenger vehicles (total)	1,305,608
up to 1000 cc displacement	318,071
1000 to 1500 cc displacement	510,207
1500 to 2000 cc displacement	258,444
2000 to 2500 cc displacement	84,266
2500 to 3000 cc displacement	51,330
3000 to 4000 cc displacement	64,681
over 4000 cc displacement	18,235

– German passenger car production during World War II (Source: Graf von Seherr-Thoss, "Die deutsche Automobilindustrie").
1940: 67,561, 1941: 35,165, 1942: 27,895, 1943: 34,478, 1944: 21,656, 1945: no data.

The wartime production of passenger vehicles thus added up to some 187,000 units, of which 51,000 were the VW Kübelwagen, 15,000 VW Schwimmwagen, 19,000 Mercedes 170 Kübelwagen, 13,000 light uniform car, 25,000 medium uniform car and 10,000 heavy uniform car. These figures are based on statistics taken from various sources, which differ to some extent.

– As for Statistics of vehicles captured or produced in occupied countries, only fragmentary statistics are available, as in the report of the GBK of 11/1/1941: "From the motor vehicle production in Germany, France, Belgium and Holland, there were produced for German use from 9/1/1939 to 9/1/1941: 128,900 passenger cars, 189,000 motorcycles."

– When Austria was occupied in 1938, 73 light and 50 heavy military passenger vehicles were taken (Source: E. Steinböck, Osterreichs mil. Potential").

– On the forced partition of Czechoslovakia into the German "Reich Protectorate of Bohemia and Moravia" and the dependent "Slovakia" in March 1939, the Wehrmacht took possession of around 1600 passenger vehicles of the Czechoslovakian Army (Source: Karl Ziskas).

– As to numbers of captured French and British passenger vehicles, unfortunately there are no statistics available.

– In Russia the Wehrmacht captured, as of the winter of 1941: 873 passenger vehicles, 32 ambulances, 4913 trucks and 573 motorcycles (Source: H. Schustereit, "Vabanque").

From the same source comes a list of losses during the first five weeks of the Russian campaign, which is thought-provoking:

Lost	9,100 motorcycles,	4772 cars,	6927 trucks;
Supplied	446 motorcycles,	337 cars,	1403 trucks;
Captured	160 motorcycles,	92 cars,	525 trucks.

The losses were considerable, as can be seen – but with the disastrous winter of 1941-42, the situation became even worse:

In the first half-year of the eastern campaign, the Army lost 103,704 unarmored motor vehicles (including 2208 trailers). The following comparison figures show the numbers of vehicles supplied (and produced) in that period of time:

Type	Lost	Gained
Motorcycles	38,601	3,073
Cars	21,559	3,089
Trucks	36,189	12,139
Tractors	2,577	276
Ambulances	1,250	360

To be sure, in 1941 there were 13,639 light, 9774 medium and 3509 heavy passenger vehicles, adding up to 26,992 units, produced for the Army and the Luftwaffe, but they did not provide the troops with the needed replacements.

It is not surprising that in this condition of need, which became even worse during the withdrawal, every private or captured completion vehicle was turned over to the troops.

The variety of types necessarily became more and more problematic. The infantry divisions in particular were equipped with more and more vehicles of questionable utility to the troops.

The armored and motorized divisions came off better in this respect, yet even in nominally fully equipped armored divisions, there were a good thirty different models of vehicles with non-interchangeable spare parts (compare von Senger and Etterlin, "Die 24. P.D.").

Passenger Vehicles in Action

On August 25, 1939, the "x-order", effective at midnight on August 26, 1939, revealed the concealed mobilization of the wartime army, initiated the advance and attack on Poland and the securing advance in the west. Within a few days the Army grew from 51 to 103 divisions – and had to be expanded by almost 3,000,000 men, around 400,000 horses and 200,000 motorized and other vehicles.

The "completed" forces of the peacetime Army possessed the following numbers of motor vehicles:
– 35 infantry divisions, first-wave, each 394 cars, 615 trucks, = 13,790 cars, 21,525 trucks
– 3 mountain divisions, each 253 cars, 618 trucks, = 759 cars, 1854 trucks
– 4 infantry divisions (mot.), each 989 cars, 1687 trucks, = 3956 cars, 6748 trucks
– 5 armored divisions, each 561 cars, 1402 trucks, = 2805 cars, 7010 trucks
– 4 light divisions, each 595 cars, 1368 trucks, = 2380 cars, 5472 trucks
– 1 cavalry brigade, 205 cars, 222 trucks, = 205 cars, 222 trucks
for a total of approximately 24,000 cars and 43,000 trucks.

"Completion" in this case usually meant using weak cadres to establish capable "back-line services" with usable vehicles of all kinds. Under these conditions, no consideration could be given to attaining uniformity of type.

As to the motor vehicle needs of Army troops and the like, and particularly the vehicles of the Luftwaffe and the Navy, there are unfortunately no statistics available.

The needs of the wartime Army in the autumn of 1939 added up to about 18,000 cars and 21,000 trucks for the newly established large units:
– 16 infantry divisions, 2nd wave, each 393 cars, 509 trucks, = 6288 cars, 8144 trucks
– 21 infantry divisions, 3rd wave, each 330 cars, 509 trucks, = 6930 cars, 5208 trucks
– 14 infantry divisions, 4th wave, each 359 cars, 536 trucks, = 5026 cars, 7504 trucks

The wartime strength specifications (K.St.N.) of German units indicate that passenger vehicles were chiefly command and communication vehicles.

On the company level, for example, the company chief, along with the company troop leader, the driver and a messenger, rode in a more or less off-road capable passenger vehicle, closely followed by motorcycle messengers, and the intelligence staff, which used passenger vehicles equipped with radio and telephone equipment. Cars were also provided for the platoon leaders and the storekeeper sergeant.

Battalion, regiment and division commanders, etc., had their own cars – and with higher rank, there was more consideration given to comfort than off-road capability in the choice of cars.

Along with the vehicles of the field troops, a goodly number of passenger vehicles (on which there are no data) were used in the hinterlands, the "homeland war zone", at Luftwaffe and naval support points and the like, in fact, wherever the fast transport of personnel, information and urgently needed goods took place.

The combat use of passenger vehicles took place at the beginning of the war, usually in the Panzerjäger companies, where the vehicles were used as towing tractors for light antitank guns and ammunition trailers. This task was taken over later by the Krupp-Protze or light towing tractors.

During the course of the war, several light fusilier regiments (such as GD) were equipped – for lack of armored personnel carriers – with heavy passenger vehicles for combat use.

After the painful experiences of the first Russian winter, the Army established "light rifle companies (mot)(VW)" for the motorcycle rifle battalions at the beginning of 1942. In March of 1943 they were renamed "armored reconnaissance companies (VW)" (of an armored reconnaissance unit). These fast-moving units were equipped with the VW Kübelwagen and Schwimmwagen, included about 190 men, 22 machine guns and two grenade launchers, and remained part of the Panzer and Panzergrenadier divisions to the end of the war (see K.St.N. 1113 PzAufkl.Kp. (Volkswg.) 11/1/1944).

Because of their meager off-road capability, and particularly because of their lack of protection from shots, the combat use of these passenger vehicles had very narrow limits.

Still in all, "the Pkw" was vital to the troops to the bitter end – and remains so to this day for all the world's armies.

Organization and Classification

Designation of passenger vehicles according to D 600 "Essential Values of Motor Vehicles and Equipment" and Wehrmacht Troop Transport Directives.

– Kfz 1: Light off-road vehicle with uniform chassis I for light passenger vehicles (l. Einheits-Pkw) or with chassis of the light Pkw (o).
Four-seat command and communication vehicles, especially also the VW Kübelwagen (l. Pkw K 1, Type 82), which was not designated as "off-road."

– Kfz 1/20: Light off-road Pkw (amphibious).
VW Schwimmwagen (l. Pkw K 2 s Type 166).

– Kfz 2: Intelligence vehicle and also radio vehicle
Three-seat vehicles on the chassis of the light uniform Pkw or l. Pkw (o), especially Mercedes 170 VK.

– Kfz 2/40: Small repair vehicle with the chassis of the light uniform Pkw or light Pkw (o).
Three-seat vehicles with a small repair shop setup in the rear.

– Kfz 3: Light survey section vehicle with the chassis of the light uniform Pkw or light Pkw (o).
Four-seat vehicles for sound- and light-measuring troops etc.

– Kfz 4: Troop anti-aircraft vehicle with the chassis of the light uniform Pkw.
Three-seat vehicle with two MG 34 in twin mount 36 for anti-aircraft defense.

– Kfz 11: Medium off-road Pkw with the chassis of the medium Pkw (o).
Four-seat Kübelwagen.

– Kfz 12: Medium off-road Pkw with uniform chassis for medium Pkw (m. uniform Pkw) and also medium off-road Pkw with towing apparatus with the chassis of the medium Pkw (o).
Four-seat Kübelwagen as command vehicle or antitank-gun towing vehicle.

- Kfz 13: Machine-gun vehicle with the chassis of the medium Pkw (o). Lightly armored machine-gun carrier.
- Kfz 14: Radio vehicle with the chassis of the medium Pkw (o). Lightly armored radio vehicle.
- Kfz 15: Telephone vehicle or
  Radio vehicle or
  Intelligence vehicle or
  Medium off-road Pkw with equipment case.
  Four-seat Kübelwagen with the chassis of the medium E-Pkw or the medium Pkw (o). "Kfz 15" was used by the troops as a collective term for all types of Kübelwagen, and even for the Steyr 1.5-ton car. Differences from Kfz 11 and 12 are recognizable only with difficulty.
- Kfz 16: Medium survey squad vehicle with the chassis of the medium uniform Pkw or medium Pkw (o).
- Kfz 16/1: Alert vehicle with the chassis of the medium uniform Pkw.
- Kfz 17: Telephone service vehicle or
  Radio vehicle or
  Cable-measuring vehicle
  with medium uniform Pkw or medium Pkw (o) with box body.
- Kfz 17/1: Radio vehicle with the chassis of the medium uniform Pkw. No external difference from Kfz 17 recognizable.
- Kfz 18: Combat vehicle with chassis of the medium Pkw (o). Two-seat support vehicle for the infantry.
- Kfz 21: Heavy off-road passenger vehicle (6-seat) with the chassis of the medium uniform Pkw or the heavy off-road Pkw (o) (8-seat).
- Kfz 23: Telephone vehicle with uniform chassis II for heavy Pkw (see uniform Pkw). Seven-seat vehicle with additional rear side door.
- Kfz 24: Amplifier vehicle with the chassis of the heavy uniform Pkw. Four-seat vehicle with box body.
- Kfz 31: Ambulance vehicle with the chassis of the heavy uniform Pkw or the light Lkw (o). Only a few Kfz 31 were built on the basis of the heavy uniform Pkw, since its own weight of 3050 kilograms was too heavy in comparison with the Phänomen Granit (2400 kg).
- Kfz 69: Limber vehicle with the chassis of the heavy uniform Pkw or the light off-road Lkw (o). Six-seat vehicle used as towing tractor for light antitank guns etc.; weight of the heavy uniform Pkw: 3150 kg, of the Krupp Protze: 2700 kg.
- Kfz 70: Personnel vehicle with the chassis of the heavy uniform Pkw or the light off-road Lkw (o). Eight- or nine-seat vehicle, with installation of an anti-aircraft machine gun possible.
- Kfz 81: Light anti-aircraft gun vehicle with the chassis of the heavy uniform Pkw or the light off-road Lkw (o). Seven-seat vehicle as towing vehicle for light anti-aircraft guns and the like.
- Kfz 83: Light searchlight vehicle I and II with the chassis of the heavy uniform Pkw or the light off-road Lkw (o). Five-seat vehicle as transport or towing vehicle for electric generators and searchlights.

Load Classes

For optimal utilization of the little space available for railroad transport, all of the Wehrmacht's rolling stock was divided into "load classes."

Light and medium passenger vehicles were in load class II, a heavy Pkw (o), for example, in load class Ib.

For the sake of thoroughness, the entire list will be given here.

Load class	Vehicle weight	Vehicle length
I	to 17.5 tons	7.21 to 11.14 meters
Ia	to 12.5 tons	6.01 to 7.20 meters
Ib	to 10.0 tons	5.01 to 6.00 meters
II	to 7.5 tons	3.54 to 5.00 meters
III	to 5.0 tons	2.51 to 3.53 meters
IV	to 4.0 tons	to 2.50 meters
A	17.5 to 26.5 tons	to 9.28 meters
S	regardless of weight	over 11.14 meters
S	over 17.5 tons	9.29 to 18.08 meters
S	over 26.5 tons	regardless of length

One R wagon of the Reichsbahn could for example, carry two vehicles of load class II.

This shows how the classification of Wehrmacht vehicles was recorded on the front doors according to regulations. Strangely enough, the net weight of this Mercedes-Benz Type 170 VK radio car is recorded as 1775 kilograms, which more or less equals its fighting weight. The vehicle's own weight (ready for action) was some 1225 kg.

Structure and vehicles of a Panzerjäger company of an infantry regiment. (from Feldgrau)

Infantry Panzerjäger Co. 1941
K.St.N. 184c

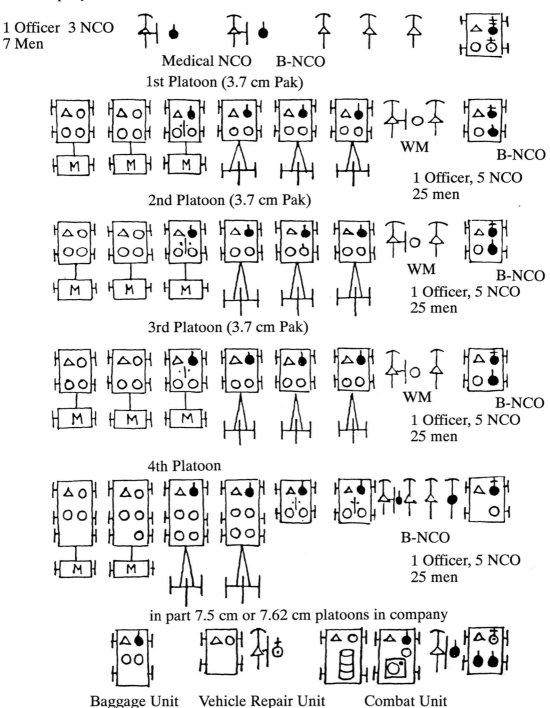

5 officers, 31 NCO, 123 men, 27 Pkw, 7 Lkw, 9 solo, 8 sidecar cycles
85 rifles, 65 pistols, 14 machine pistols, 5 light MG, 9 3.7-cm Pak, 2 5-cm Pak

Structure and vehicles of an armored observation battery. The numbers under the vehicles give their vehicle numbers. (from Feldgrau)

Panzer Observation Battery 1940-1943

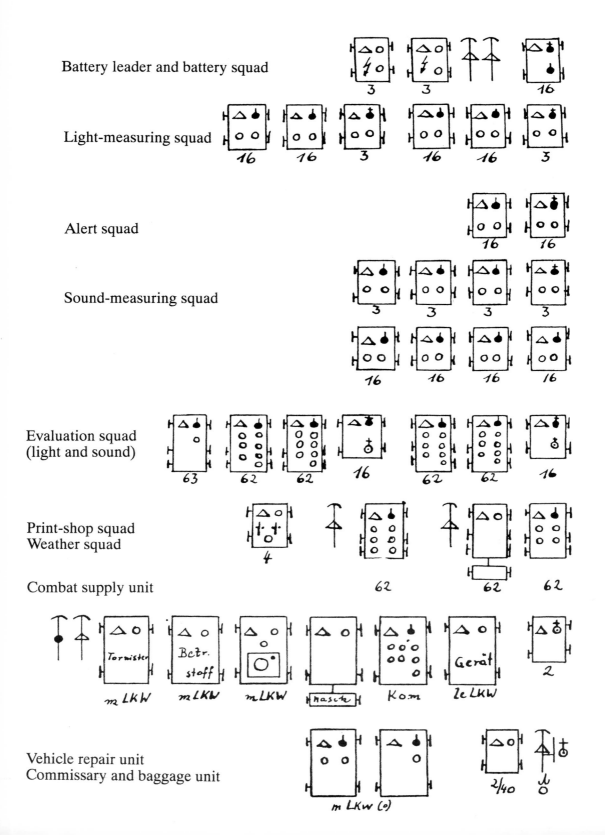

Structure and vehicles of an Army anti-aircraft company. The drawing is from "Hermann Freter-Fla forward!" and shows clearly which soldiers were assigned to the individual vehicles.

AA COMPANY 1944

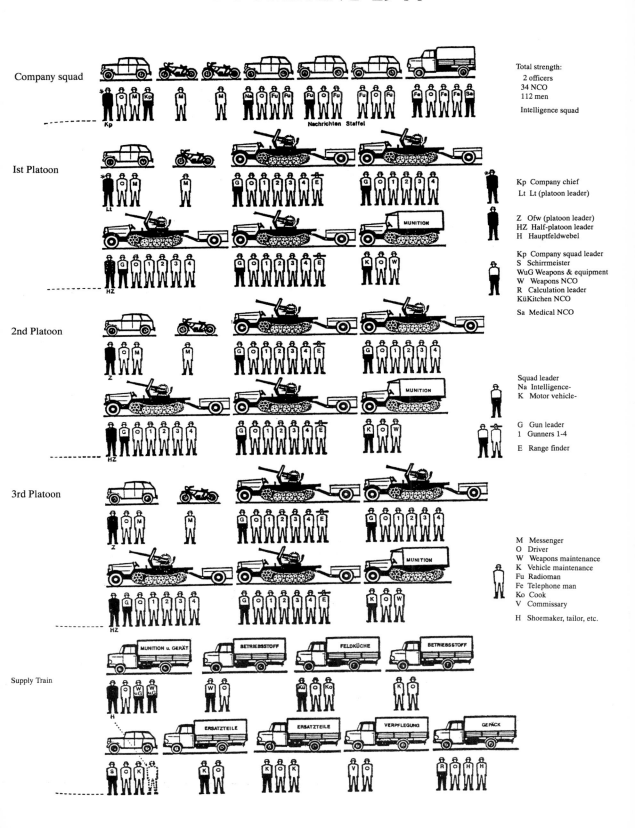

Makes of Cars in Germany

The following list comes from the materials used in troop engineer training in 1944. It shows, on the subject of anti-freeze, the contents of the cooling systems of the makes and models of cars found in Germany (circa 1937). This list is certainly not absolutely complete, but it gives a good overview of the passenger cars that were then current in Germany – and surely requisitioned.

Make	Model	l.	Make	Model	l.	Make	Model	l.
Adler	Trumpf Junior 1E, 1G	9		Mod. 50 1934-35	15		Sp. 8, Sp. 9	26
	Trumpf 1.5 AV, 1.7 AV, 1.7 EV	11	Buick	Model 60 1934-35	17	Hanomag	2/10	6
	Primus 1.5 A, 1.7 A	9		Model 90 1934-35	22		16, 18, 20, 23	9
	Primus 1.7 E	11		Model 40 1936	13		6/32	10
	Favorit 8J, 2S, 2A, 2U	15		Model 60, 80, 90 1936	16		Kurier 4/23	11
	Diplomat 3 G	18	Cadillac	V 8 1931-33	23		Record 6/32	12
	Diplomat 3 G d	20		V 8 1934	19		Sturm 9/50	13
	Standard 6 A, 6 S, 6/3U	16		V 8 Series 60 1936	28	Hansa	1100	8
	Standard 6	17		V 8 Series 70, 75 1936	27		1700, 2000	10
	Standard 8	22		V 12 1931-33	25		3500	16
	2.5 L 58 PS	15		V 12 1934	17	Hispano-Suiza	25/100	27
Alfa Romeo	6 C 1500, Sport, Normale	10		V 12 Series 80,85 1936	18	Horch	420, 450, 500 B 20/100	24
	6 C 1750, Turismo, Gran Turismo	10		V 16 1931-34	22		600/670 12 Cyl.	34
	Gran Sport	12	Chevrolet	V 16 Series 90 1936	23		750 B, 780 20/100	24
	6 C: Gran Turismo, Supercharged	12		Standard Model EC	10		850, 851, 853	28
	6 C 2300: Turismo, Gran Turismo, Pescara	14		Master Model EA	10		830 Bk, 830 Bl, 830 V	21
	6 C 2300 B	13	Chrysler	All series 1936	12		951	25
	8 C 2300: Monza, Gran Sport, Tipo Lungo	15		Six 1931-34	15	Lancia	Aprilia	7
	8 C 2900 B	13		Eight 1932-33	19		Augusta	8
Amilcar	4/20, 5/30	8		Eight 1934	22		Artena	13
	10/65	27		Imperial 1933	20		Lambda 8th & 9th Series	14
Auburn	6/80-6/85	16		Imperial 1934	22		Astura 3rd Series	17
	8/90-8/95, 120-125	19	Citroen	Mod. 1930-31 C4F	10		Dilambda, Dilambda-Sport	27
	8/98-8/98 A	24		Mod. 1932 C4G	12	La Salle	1932-33 V 8	23
	8/100, L 29	22		Mod. 1933-34 8A	10		1934 8	17
	12/160	42		Mod. 1933-34 10A	12		1936 36-50	15
Audi	P 530	11		Mod. 1934-36 Types 7 & 11	8	Martini		25
	14/50	32		Mod. 1931-32 C6F	14	Mathis	EMY-4, M4s	12
	Dresden 3.9, 15/75	23		Mod. 1932-33 C6G	14		FO4	14
	UW 6 Cyl. Front, Type 225	14		Mod. 1933-34 15A	17		EMY-6, FOS	14
Austin	3/15, 3/17	6	Daimler-Benz	see Mercedes-Benz		Maybach	DS 12 Cyl. 150 HP	ca. 43
	5/25, 5/28	10	De Soto	Six 1931-33	14		DS 12 Cyl. 200 HP	ca. 43
	6/30, 6/35	13		Six 1933, Eight	15		W 3/6 70 HP	ca. 24
	7/40	14		Six 1934	19		W 5/6 120 HP, W 6/6 120 HP	ca. 26
Austro-Daimler	ADM II, ADR Sport	20	DKW	Frontwagen models	8		DSh/6 130 HP	ca. 26
	ADR	21		Schwebeklasse	11		3.8 Ltr. 6 Cyl. Swing Axle	ca. 14
	ADR Bergmeister	16		Sonderklasse	12	Mercedes-Benz	130	10
	ADR 8 Cyl.	25	Essex	1927-32	21		150	13
BMW	0.9 Ltr., 1.5 Ltr., 1.5 Ltr. Sport	7		Terraplane 1932	12		170, 170V, 200, 230, 260 D, L 300	11
	2 Ltr., 2 Ltr. Sport	8		6 C 1933, 1937	12		170 H	9
	2 Ltr. 1938 Type 328	8		8 C 1933	15		Stuttg. 200, 260, L 1000	15
	2 Ltr. Sport Type 328	7	Fiat	8 C 1934-35	17		290, 350, 14/60	19
Brennabor	C 4/20, D 4/22	10		6 1936	13		320 (WO 4) 1926-28	18
	Ideal-Extra 7/30	12		Hudson 6	17		320 1937	12
	E 8/38	16		Hudson 8	22		370, 370 K, 370 S	18
Bugatti	Type 40 1.5 Ltr.	10		Hudson 6 1936	13		380	20
	Type 44 3 Ltr.	11		Hudson 8 1936	20		480, 500, 500 K, 540 K	26
	Type 46 5.3 Ltr.	23		Hudson 6 1937	12		770 Gr. Mercedes	32
	Type 49 3.3 Ltr.	12		Hudson 8 1937	19	Nash	480	15
	Type 50 4.9 Ltr.	23		500	5		Nash Adv. 6	16
	Type 55 2.3 Ltr.	10		508 "Balilla"	7		490	20
	Type 57 3.3 Ltr.	12		514, 515, 1500	8		450, 660, 660 KB	12
Buick	8/50 1931-32	12		509, Ardita	10		870	13
	8/60 1931-32	14		520, 520 T	13		880	16
	8/90 1931-32	18		521, 521 C	14		890	21
	32-50, 33-50	12		522-524	13		1130/80	15
	32-60, 33-60	15		525, 525 N, S, SS, NS	19		Big six	17
	33-80/90, 32/90	18	Ford	527 (Ardita 2500)	12	NSU-Fiat	Mod. 1000 4/25 HP	7
	Mod. 40 1934-35	13		Model A, AF	11		Mod. 1500 6 Cyl.	8
				Model B, Rheinl.	13	Opel	8/40 1929-30	17
				Y 4/21 PS Köln	7		4/20 1930	12
				C 5/34 PS Eifel	7			
				48	21			
				51 V 8	24			
			Gräf & Stift	MF 6	17			
				Sp. 5, Sp. 6, G 8, G 35, G 36	25			

Make	Model	l.	Make	Model	l.	Make	Model	l.
Opel	1.1 Ltr. 1931	12		Code PJ	15		S 8 8/45	16
	1.2 Ltr. 1932-35	6	Renault	Celtavier, Monavier	12		S 10 10/50	19
	1.8 Ltr. 1931-33	8		Primavier, Viravier	13		V5 5/25 front drive	12
	1 Ltr. 1933	9		Primastella	16		V 8 front drive	20
	1.3 Ltr. 1934-35	8		Vivasport, Vivastella,			R 140/30 front drive	10
	1.3 Ltr. Olympia	7		Viva Grand Sport	22		R 150/32 front drive	10
	1.1 Ltr. P4 1935-36 to			Nervasport, Nervastella,			R 180/42 front drive	12
	#33000	9		Nerva Grand Sport	22		Sedina	18
	1.1 Ltr. P4 1936-37 from			Reinasport	24		Arkona	21
	#33001	6	Röhr	9/50	12	Studebaker	Six, Dictator	14
	1.1 Ltr. Kadett 1937	6		RA 2.5 Ltr. 10/55	20		Commander 8	18
	2.0 Ltr. 1934-37	9		F 3.3 Ltr. 13/75	25		President 8	18
	2.5 Ltr. Super 6	12	Rolls-Royce	17/80	23	Tatra	23/24	36
Packard	120	16		30/150	31		24/58	47
	326-33	18	Standard	all types	8		27	26
	726, 733	19	Steyr	30, 30 S, 30 E, 430, 45, 50	10		70	30
	740	24		100, 120 S, 200, 220 S	11		80	35
	745	25		530, 630	12	Wanderer	6/30	8
	833, Light Eight, 901, 902	20		440	20		7/35, 8/40	7
	1002	19	Stoewer	D 9 9/32, D 9 V 9/38	19		10/50, 12/65	11
	Twin-Six	40		F 6 6/30	10		W 235, 240, 250	8
Peugeot	201	9		614 14/70, 614 K 14/70	19		W 23, 26, 52	10
	301	10		615 15/80, 615 K 15/80	23		W 24	7
	401	11		M 12 12/60	20		W 25 Sport	8
	402, 601	12		P 20 20/100	36		W 35, 40, 50, 51	8
Plymouth	- -	14						

Source: "Schlag nach!", 1940 edition

German Cars

Uniform Passenger Vehicles

The Light Uniform Pkw

In D 662/1 of April 22, 1940, it is stated: "The uniform chassis I for light Pkw is an all-wheel-drive Kfz of the Army's own type. The construction of the chassis is uniform except for the choice of installing various motors, lubrication systems and electric equipment by any manufacturers."

The first version, which was built from 1936 to 1940, had four-wheel steering. This did not work out, though, since it led to dangerous handling problems at speeds over 25 kph and offered no definite advantages in off-road terrain. Four different motors, with attachment points and crankshafts made to be uniform, were installed:

1. Hanomag Type 20 B, 2 liters (1991 cc, 4 cylinders, 48 HP: sustained performance)
2. BWM Type 325, 2 liters (1957 cc, 6 cylinders, 45 hp sustained: performance)
3. Stoewer Type R 180W, 1.8 liters (1757 cc, 4 cylinders, 42 HP: sustained performance)
4. Stoewer Type AW 2, 2 liters (1997 cc, 4 cylinders, 48 HP: sustained performance)

The exchange of motors involved considerable difficulties in making them fit. The Stoewer 1.8 liter, for example, had the fuel intake on the right and the exhaust manifold on the left – the exact opposite of the 2-liter Stoewer. There were two different instrument panels, depending on whether the electric uniform switch box was installed or not. Lubrication was done by hand in the first series, with 42 lubrication points on the steering the drive train that had to have grease applied every 1000 kilometers. For cars with central lubrication, it was only necessary "to step on the pump pedal briefly and strongly once" every 50 to 100 kilometers to press motor oil to the lubrication points through a system of pipes.

These non-uniform features of the uniform vehicle were supposed to be eliminated in 1940 by the "Uniform Chassis I for light Pkw Type 40." The "Type 40" had the two-liter Stoewer AW 2 motor, two-wheel steering, oil-pressure instead of cable brakes, the uniform switch box, central lubrication, and a chassis weight lightened by 100 kilograms to 1280 kg. The production of the chassis and body was now carried out exclusively by the firm of Stoewer in Stettin. The previous chassis had been made not only by Stoewer, but also by BMW in Eisenach and Hanomag in Hannover, and the bodies were generally made by Ambi-Budd in Berlin.

In action, especially under "Russian" conditions, severe defects in the frame, wheel suspension, clutch, driveshaft, steering, etc., became apparent. But despite all of these faults, many former soldiers – with a lot of hindsight – still fondly remember this extraordinarily off-road capable and nimble vehicle.

"Uniform Chassis I for light Passenger Vehicles" with four-wheel steering. At the rear is the main with fuel tank 50-liter capacity; an additional container with 10-liter capacity was attached to the bulkhead. This fuel supply, according to 600 D, was enough for a range of 350 km on the road and 240 km in medium terrain. The chassis shown here has a Stoewer 1.8 liter Type R 180 W motor.

Instrument panel of the Uniform Chassis I for light Pkw (with uniform switch box)m according to D 662/1 of 4/22/1940.

1. Oil pressure gauge
2. Water thermometer
3. Dashboard light
4. High beam indicator light
5. Speedometer
6. Starter button
7. Hand throttle
8. Flasher switch
9. Wiper switch
10. Uniform switch box
11. Light switch
12. Cooling flap lever
13. Clock
14. Fuel stopcock
15. Attaching plate
16. Central lubrication control

Two radio cars (Kfz 2) of Armored Engineer Battalion 79 of the 4th Panzer Division on the way to Beresina. The Kfz 2 was equipped with a b1 or f canister radio device. The unit consisted of the devices and the accessories and battery container. These were attached by removable elastic straps to the base of the container next to the left rear seat. During transit, radio communication could be carried on using either the generator antenna or the vehicle's removable two-meter antenna. In all light uniform vehicles with a vertical rear, the spare wheel was stowed in the rear above the main fuel tank. No place for personal equipment, clothing, food, bedroll, etc., was provided in this car, so everything had to be loaded in sacks and boxes 'somewhere.'

Two "Radio Car (Kfz 2) with the Uniform Chassis I for light Pkw" of the battalion staff of Engineer Battalion 18 just outside Vilna in the summer of 1941. In the background are radio cars of the intelligence platoon. The antenna socket for the vehicle rod antenna can be seen on the left side of the car.

Numerous bullet holes in the right (always one-door) side of this Kfz 2 of AA Battalion 47 show the vulnerability of unarmored vehicles in action. The "Fla" was the anti-aircraft weapon of the Army, while the "Flak" was that of the Luftwaffe. This picture was taken in Albania or Greece in the spring of 1941.

A light uniform Pkw of an unknown Luftwaffe unit at a facility in Smolensk. By the vertical rear and the lack of an antenna, this could be either an "Intelligence Vehicle (Kfz 2)" or a "Small Repair Car (Kfz 2/40)."

Two light and one medium uniform vehicle near the Düna at Jakobstadt at the beginning of July 1941. The medium uniform Pkw (obviously a Kfz 15) belonged to Panzer Regiment 1 and bore the flag of a staff unit; on the license plate (!), the symbol of the 1st Panzer Division is visible. Behind it are a motorcycle messenger and a "Radio Car (Kfz 2) with the Uniform Chassis I for light Pkw." The vehicle passing them is a light uniform Pkw of the Luftwaffe.

The "Troop Anti-Aircraft Gun Car (Kfz 4) with Uniform Chassis I for light Pkw" was very seldom seen. According to war strength information, more than a single Kfz 4 was never planned, and it was used, for example, by the gun crew of a light howitzer battery or the evaluation platoon of a survey battery. The vehicle was much too small for two MG 34 on a Type 36 twin mount and could obviously fire only forward. In the "Description of the MG 34 . . . in Anti-Aircraft Defense" by Oberst A. Butz of May 1943, it is no longer mentioned. The Kfz 4 had a straight rear with external spare wheel, and probably only three doors.

A "Light Off-Road Passenger Car (Kfz 1)" or "Light Survey Troop Car (Kfz 3)" of the Hungarian Army. Both types of vehicle had a straight rear with external spare wheel, a four-man crew and four doors.

These "Panzerkampfwagen Imitations" on the "Uniform Chassis I for light Pkw" were used by I.R. 4 at the Hindenburg Barracks in Glogau in the autumn of 1940. The wooden body was removable. At left is a Wanderer W 14, being entered by the regimental commander and bearer of the Knight's Cross, Oberst Recknagel (who was KIA as commanding general of an army corps on January 18, 1945).

Willy, the "Tommy with the French helmet", is "disposed of." At that time a landing in England was still under consideration – and nobody was thinking of a war with Russia. On the left fender, the emblem of the 18th I.D. (mot.) can be seen, along with the covered shell of the headlight. I.R. 54 became the 100th Light Infantry Division on December 10, 1940.

A particular rarity among the light uniform cars was the narrow "Mountain Hunter Version" with its track decreased from 1400 to 1250 mm. This vehicle was photographed on the road from Trondheim to Kirkenes in 1942. It bears the tactical symbol of the 2nd battalion and the yellow Edelweiss of the 6th Mountain Division.

A light uniform Pkw in "Mountain Hunter Version" amid the depressing scene of a shot-up French column near Mondrepuis. The picture comes from an album of the II./Geb.Art.Rgt. 79.

The Medium Uniform Pkw.

The "Uniform Chassis for medium passenger vehicles" existed in the version "with support axle", the "Type 40" and "with cabriolet body." All of these vehicles had four-wheel drive and front-wheel steering.

Production of these vehicles was carried out by the Auto Union AG and the Adam Opel AG.

The following chassis serial numbers are listed in D 663 ff:

No. 10,001 ff: Auto Union, Wanderer works, Siegmar, "support axle", Horch V-8 cyl., 3.5 liters, 80 HP.
No. 20,001 ff: Auto Union, Horch works, Zwickau, otherwise as above.
No. 30,001 ff: Opel, Rüsselsheim, Brandenburg, "support axle", Opel 6 cyl. in line, 3.6 liters, 68 HP.
No. 40,001 ff: Auto Union, Wanderer works, Suegmar, "Type 40", Horch V-8 cyl., 3.5 liters, 80 HP.
No. 50,001 ff: a. "Type 40", Horch works, as above. b. "cabriolet body", Horch works, Horch V-8 cyl., 3.8 liters, 90 HP.

The "Uniform Chassis Type 40" differed externally only in its lack of lateral support wheels and its lengthened fuel fillers.

The replacement of a Horch motor with an Opel motor was done by modifying the fuel and electric lines, engine protection panels, radiator supports and the like, replacing the clutch including the rods, sometimes installing a new exhaust system, new longitudinal suspension members, axle ratio and auxiliary gears and, depending on the type, a new gearbox.

The medium uniform Pkw also showed serious defects in its frame, wheel suspension and power transmission when in action. The D 663/1 (of 1/16/1942) thus included a section on strengthening the axles and installing additional springs in the field.

The vehicles with support axles were only partially equipped with central lubrication, while all the Type 40 vehicles had it. The off-road capability was strongly limited by the meager ground clearance and the high fighting weight (Fu. Kw. Kfz 15, circa 3.3 tons). But the motors that were used were highly praised for their reliability and power, even though they caused many servicing problems.

1. Towing hitch
2. Tailpipe
3. Fuel tank float
4. Reserve fuel tank
5. Fuel filler cap
6. Upper link
7. Spring cups
8. Half-shaft
9. Rear axle drive
10. Drive wheel
11. Shock absorber
12. Hydraulic brake line
13. Brake cable
14. Main fuel tank
15. Battery
16. Driveshaft (to rear wheels)
17. Spare wheel mount
18. Spare wheel
19. Differential gearshift
20. Transmission gearshift
21. Hand brake lever
22. Brake pedal
23. Clutch pedal
24. Steering wheel
25. Right rear brake line
26. Differential air valve
27. Differential
28. Hand brake shaft
29. Driveshaft (from transmission)
30. Hand brake lock
31. Transmission
32. Driveshaft (to front axle)
33. Air cleaner
34. Coil
35. Distributor
36. Fuel pump
37. Intake tube
38. Muffler
39. Water pipe with thermometer
40. Radiator hose
41. Radiator
42. Ventilator
43. Oil filter
44. Tow hook
45. Generator
46. Brake line divider

Chassis of the medium off-road uniform Pkw with towing axle.

"Medium Off-road Pkw with Uniform Chassis for m. Pkw" of the Luftwaffe. According to HDv 68/5 (as of 4/1/943), there were a "m. gl. Pkw (Kfz 12)" and a "m. gl. Pkw with Toolbox (Kfz 15)." No external difference can be seen. The troops spoke almost exclusively of Kfz 15. In the boxes mounted on both sides behind the front fenders, the tire chains were kept. Note the ventilator flaps – in place of louvers – on the hood of the vehicle shown here.

The interior of a medium uniform Pkw of Rifle Regiment 394.

Two medium uniform Pkw of Intelligence Unit 27 during prewar maneuvers at Mühldorf on the Inn.

The company squad of the 5th Company of Railroad Engineer Regiment 3 scouting near Bastogne. The cable rack on the fender indicates that this may be an "Intelligence Vehicle (Kfz 15) with Uniform Chassis for medium Pkw", used here as a survey car.

A medium uniform Pkw (WH-242 853) of the 2nd/Engineer Battalion (mot.) 20 (Hamburg) drives around a Czech roadblock during the advance into the Sudetenland in October 1938. On the rear is a small raft sack, on the motor hood a paddle, and behind the driver a roll of wire. This doorless vehicle of the 1st Series had a head-on collision with a uniform Diesel in Paris in the summer of 1940 which put it totally out of commission.

A Medium Uniform Pkw of Engineer Battalion 156 crosses the Ysere on a float ferry.

Soldiers of one of the five Army Panzerjäger units standing by with the "4.7 cm Pak (t) (Sf) on Panzer 1 chassis." At left is an intelligence car (Kfz 15) with markings visible from the air, behind it a requisitioned Mercedes and a truck resembling the "Granit."

A medium uniform Pkw with its tailgate open in the Ilmensee area.

The 16th (Panzerjäger) Company of Mountain Jäger Regiment 85 near Atlanthi, Greece, at the Channel of Atlanthi, south of Thermopylae.

At left is the company leader with the company squad and then the four platoons, equipped with 1-ton towing cars for 3.7 cm Pak guns, with the platoon leader with a Kfz 15 at the head of each row. At right is the baggage and supply train with uniform Diesels, and on the outside the Kfz 2/40 of the auto repair squad.

In the mud of the Ukraine, fenders were of great importance; otherwise, as here, the handsome vehicle would be plastered with dirt until it was unrecognizable. This shows a medium uniform Pkw of the "Grossdeutschland" I.D. (mot.).

Vehicles of the 1st Medical Company of the 6th SS Mountain Division "Nord" resting in Sodankylä, Finland, in the autumn of 1942 (SS-97 270).

A good view of the medium uniform Pkw from the front, this time in a ditch on the road to Nivalla. This road apparently invited such accidents (see "Ford in ditch" in the book "Trucks of the Wehrmacht", page 80!)

The last battle in the western campaign for Panzerjäger Unit 31 took place on June 19, 1940 at Chateau Gontier (west of Le Mans). The Germans lost two scout cars, two Kübelwagen and one motorcycle. This picture shows a burning medium uniform Pkw of the 1st Platoon (WH-69714).

This photo from an album of the II./Mountain Artillery Regiment 79 is captioned: "This Pkw (medium uniform Pkw) of the 16th Company drove over a Russian mine in the attack on Prigoshayu (5/21/1942)." Note the depth of the crater.

The medium off-road passenger vehicle with "Uniform Chassis for medium Pkw Type 40" for the commander of Engineer Battalion 18, Oberstleutnant Schmeling (right front) near Dashki in June 1941. Standing is the battalion adjutant, Leutnant Melzer. The "Type 40" differed from the type "with towing axle" in particular by its widened passenger space and the lack of the towing axles on both sides. Just one spare wheel was carried (inside on the left). The filler for the centrally located main fuel tank was also relocated to the outside, so that filling no longer had to be done with a can inside the car. The fillers for the main (70 liter) and auxiliary (40 liter) fuel tanks were now in a niche that had to be built into the bodywork.

The medium uniform Pkw "Type 40" (WL-13 088) of the staff company, III/JG 52 during transfer from Nikolaiev to Kerch in March 1943. At left is Oberleutnant Schwinn, chief of the staff company, at right Leutnant Lamprecht; the driver is unidentified. The inside spare wheel is easy to see.

Vehicles of the "Grossdeutschland" Panzer Grenadier Division in the Tomarovka area in the spring of 1943. In front is a "Radio car (Kfz 15) with uniform chassis for medium Pkw (Type 40)." The tactical symbol under the division's emblem is that of the staff of a motorized artillery regiment. Behind it is a "Radio car (Kfz 2) with uniform chassis I for light Pkw", followed by two radio cars (Kfz 17/1) with rod antennas and two radio cars (Kfz 17) with frame antennas. The Kfz 17 with uniform chassis for medium Pkw had four seats and a fighting weight of over 3.6 tons.

Communication activity by an unidentified unit (the photographer was formerly with Motorcycle Battalion 402) in occupied France, 1943-44. The Kfz 15 probably has dark yellow paint with olive green and red-brown camouflage. The Kfz 17 behind it, with its frame antenna up, seems to be painted dark yellow, visible from far away.

The later version of the Radio Car Kaf 15 had the chassis of the medium uniform Pkw Type 40 and can be recognized by the extended fuel fillers and the lack of spare wheels.

Rear-guard action of IR 61 near Novgorod Seversky on 9/5/1943. This medium uniform Pkw (WH-62 488) carries the standard of the division commander and bears the emblem of the 7th ID and the staff pennant on its fender.

Observing the enemy before Voronesh in July 1942: This "medium survey car (Kfz 16) with uniform chassis for medium Pkw (Type 40)" of the 1st Battery of the "Grossdeutschland" Artillery Regiment has come up to join the tanks. In the background is the first "long-barreled" version of the Panzer IV, the F 2 type, of which 200 reached the troops beginning on March 1942.

Tunisia, 1943: A medium uniform Pkw Type 40 of JG 77 serves as a buffet for a hasty wartime meal before returning to duty.

For high-ranking troop leaders, a special commander's car was created at the beginning of the war. According to D 663/11 (9/22/1942), it was designated "Uniform Chassis for Medium Pkw with Cabriolet Body (without towing axle, with Horch 3.8 liter motor)." This was the Army's own special version of the uniform chassis for medium Pkw, Type 40, which was used only with the cabriolet body. The major changes were to the motor (now 3.8 liters, 90 HP) and the transmission and reduction gears. – In this picture, the commander of the 10th Panzer Division, Generalmajor Fischer, and his cabriolet are seen with Rifle Regiment 69 in the Selivan-Lenino area, 25 kilometers northwest of Moscow. This front line had to be evacuated suddenly on 12/8/1941 at -35 degrees, with heavy losses – not surprising with the fully unsuitable winter equipment of the German troops.

General Traut sees off Commanding General Henrici in his command cabriolet.

The Heavy Uniform PKW

The four-wheel-drive "Uniform Chassis II for heavy Pkw" was built by the Auto-Union AG at its Horch works in Zwickau and the Fordwerke AG in Köln-Niehl and Berlin. On the basis of the chassis numbers listed in D 664 ff, some order may be brought to the colorful variety of various types:

– No. 100,001-100,350: "Type a" 4-wheel steering (to max. 25 kph) shifting to 2-wheel steering, divided Horch transverse links, mechanical brakes, Horch 3.5 liter V-8 motor. 80 HP.
– No. 100,351 ff: "Type 1a", 4-wheel steering, shifting to 2-wheel steering, Rheinmetall transverse links, oil-pressure brakes, Horch 3.8 liter V-8 motor, 81 HP.
– No. 120,001-120,650: "Type b", 2-wheel steering, otherwise as Type a.
– No. 120,651 ff: "Type 1b", 2-wheel steering, otherwise as Type 1a.
– No. 140,001 ff: "Type 1c", 2-wheel steering, only for bodies with side armor (light armored scout cars with rear engine).

In these vehicles, the fillers for the main and auxiliary fuel tanks were internal. All vehicles had support axles (except Type 1c) and were built by Horch.

Numbers 300,001 ff were built by Ford, as follows:
– "Type EGa", 4-wheel steering (to max. 25 kph), shifting to 2- wheel steering, oil-pressure brakes, support axles, Ford V8 3.6 liter motor, 78 HP, fuel fillers extended to the side.
– "Type EGb", as above, but two-wheel steering.
– "Type EGd", as Type EGa, but without support axle.

Numbers 400,001 ff: "Type 40", made by Horch, with Horch V8 3.8 liter motor, 90 HP.
Numbers 500,001 ff: "Type 40", made by Ford, with Ford 3.6 liter V8 motor, 78 HP.
All the models had a central lubrication system. Depending on their use, 130, 300 or 600 watt generators were installed. In the "Type 40", all-wheel steering and support axles were eliminated, and the fuel fillers were extended to the side.
While in action, the vehicles often suffered frame and spring breakage. An addition to D 664/7, "axle strengthening and installation of auxiliary springs", therefore appeared, for the purpose of avoiding additional damage. The heavy uniform Pkw was very popular among the troops for its roominess. It was intended for the same uses as the three-axle Krupp L 2 H143 light off-road truck (o). Like all uniform passenger vehicles, though, the heavy uniform Pkw suffered from its complicated design and its heavy weight. The ready-to-use inherent weight of the Kfz 70 with the heavy uniform chassis was 3150 kilograms, but only 2485 kg with the Steyr 1500 A chassis! The Steyr 1500 A light truck and the Mercedes L 1500 A replaced the heavy uniform Pkw as of 1941.

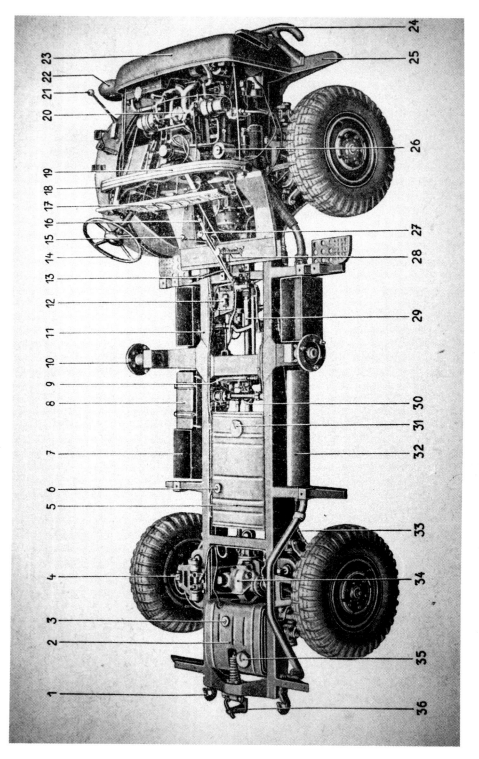

The Uniform Chassis II for the heavy Pkw Types a/b, 1a/1b/1c, according to D 664/303 of 12/22/1943 (reprinted October 1944).

1. Rear tow hook, 2. Main fuel tank, 3. Fuel gauge float, 4. Rear swing axle, 5. Reserve fuel tank, 6. Fuel gauge float, 7. Tire chain box, 8. Battery, 9. Reduction gears, 10. Spare wheel mount, 11. Longitudinal link, 12. Two-step brake cylinder, 13. Shift lever, 14. Steering wheel, 15. Shift lever, 16. Hand brake, 17. Instrument panel, 18. Front fender, 19. Firewall, 20. Motor, 21. Flagstaff, 22. Headlight, 23. Radiator cowling, 24. Bumper, 25. Transverse link, 26. Lubrication oil tank, 27. Fuel valve, 28. Oil pump lever, 29. Front longitudinal link, 30. Rear longitudinal link, 31. Reserve fuel filler, 32. Muffler, 33. Hear brake cable, 34. Rear drive gears, 35. Main fuel filler, 36. Trailer hitch.

Dashboard and controls of the Uniform Chassis II for heavy Pkw, Type EG a/b and /d, with 3.6 liter Ford motor, according to D 664/1 of 7/29/1940.

1. Width indicator
2. Ignition key
3. Starter button
4. Cooling flap lever
5. Wheel position indicator
6. Horn button
7. 2/4wd indicator
8. Oil pressure gauge
9. High beam indicator
10. Directional light lever
11. Water & oil thermometer
12. Electric plug
13. Throttle
14. Fuel gauge
15. Speedometer
16. Dashboard light switch
17. Clock
18. Ventilator
19. Gear indicator
20. Lubrication diagram
21. Clutch pedal
22. Brake pedal
23. Dimmer switch
24. Gas pedal
25. Steering lock
26. Fuel switcher
27. 2/4wd locking lever
28. Gearshift lever
29. Hand brake
30. Lubrication control
31. Jack
32. Jack handle

Heavy uniform Pkw's stand ready at Preussisch-Eylau, East Prussia, shortly before the Russian campaign began.

"Strays at Tamsaln, 8/7/1941" – This picture of Bicycle Battalion 402 shows the roominess of the "Personnel Car with Uniform Chassis II for Heavy Pkw", which was intended to carry eight men including the driver.

At least Bicycle Battalion 402 was officially equipped with towing facilities for ten bicyclists (a squad). In this picture, a heavy uniform Pkw of this battalion is seen in action in France; such action naturally required appropriate roads.

Antwerp, autumn 1940: Loading the 1st Mountain Division for the planned landing in England. Left: Uniform Diesel, Opel Blitz; right: three heavy uniform Pkw and a BMW sedan.

The staff of Mountain Panzerjäger Unit 44 studies the map at the beginning of the Russian campaign. The heavy uniform Pkw of this unit had masked headlights mounted on the radiator shell.

A heavy uniform Pkw of the 2nd Squad of Fighter Squadron 1 (WL-85 673). Note the pierced wheels.

An interesting special design built in small numbers was the "2 cm Flak (Sf) on Heavy Uniform Pkw." Here is one of those self-propelled gun units of the 7th Battery of the "General Göring" Regiment. This unit was part of the "Führer Escort Command" and had the job of protecting the Führer's headquarters from attack by low-flying planes.

This self-propelled gun unit with a shield for the 2 cm Flak gun, towing a Sd Ah 51, belonged to the 7th or 8th RGG. Besides the version shown here, with folding gun platform, there were also heavy uniform Pkw with Flak guns mounted directly on the standard body. According to Appendix 1 to D 664/1 and /7 of 2/4/1943 for "normal" heavy uniform Pkw, ". . . in all vehicles, axle strengthening and additional springs are to be installed . . .", so it appears that the chassis and frames of these vehicles were often quite overloaded.

The "Telephone Car (Kfz 23) with Uniform chassis II for Heavy Pkw" was intended to carry of crew of seven and had six side doors. The vehicle shown here (according to D 600) is notable for its horizontal engine louvers and the small number of them on the top of the hood. Note the removed mirror on the rear and the peg on the front of the rear fender.

The telephone car (Kfz 23) of an unknown unit in southern Russia. Unfortunately, the photo does not show the entire rear. The length (4850 mm) and the wheelbase (3000 mm) of the vehicle are cited in the available technical data as identical to the four-door basic version, although certain doubts arise here.

As of 1940, the "Uniform Chassis II for Heavy Pkw, Type 40" appeared, with widened body, inside spare wheels and extended fuel fillers. This model was built by both Horch and Ford. The 3.8 liter, 90 HP Horch motor had an average consumption of 30.5 liters, the 3.6 liter, 78 HP Ford of 35.0 liters.
This picture shows a vehicle of the 3rd SS Panzer Division "Totenkopf" south of the Ilmensee in the winter of 1941-42.

A bogged-down heavy uniform Pkw Type 40 of an unknown telephone unit.

The "Light Snowplow Type E" on the heavy uniform Pkw (WH-97 574), "E" standing for "one-sided." The angle of the plow could be set by a steering rod. Note the lack of any weather protection.

There were also small numbers of a command car version of the heavy uniform Pkw. This "Uniform Chassis II for Heavy Pkw Type 40 with Cabriolet Body" was obviously built by Horch, with the modifications to the drive gears of 4.375 (instead of 5.14) listed in D 664/7 of 9/15/1941 – without further information. This made a top speed of 90 kph (instead of 76 kph) possible, but the off-road climbing ability dropped from 54.6% to 46.2%. The 3.8 liter, 90 HP Horch motor was also the same as was used in the medium uniform Pkw with cabriolet body. In this picture, the commander of the "Hermann Göring" Panzer Division, oak-leaf bearer Paul Conrath, is shown at the Grafenwöhr training camp in the summer of 1943, inspecting the newly established armored artillery regiment.

German Kübelwagen

The Adler bucket-seat car based on the "Favorit" was the first off-road passenger vehicle, along with the Daimler-Benz "Stuttgart 2000", built for the Reichswehr in 1927-28. This is one of the first models, with the spare wheels still carried on the rear of the vehicle. The typical "bucket seats" are easy to see in this picture.

An Adler Type 12 N Kübelwagen with body by the Kathe firm of Halle-Saale, made in 1934. The chassis was based on the Adler Standard 6 and had a six-cylinder, 2.9 liter, 50 HP motor.

An Adler Type 12 N of Mountain Artillery Regiment 79 crossing the Oisne-Aisne Canal along with beasts of burden of the 1st Mountain Division. This Adler model was particularly similar to the Wanderer W 11 and other Kübelwagen of the 1930s.

A Kübelwagen with equipment box was built on the standard chassis of the Adler Trumpf (front-wheel drive, first built in 1932). Here one of these very rare vehicles is seen at an army driving school.

The Adler Type 3 Gd was built from 1930 to 1940, about 4300 in all. It had rear-wheel drive and the same motor as the Type 12 N-3G (6 cylinders, 2.9 liters, 60 hP). The vehicle shown here was hit by a Junkers W 34 at Peenemünde on March 14, 1939.

The Adler Type 3 Gd of a commanding general (unfortunately unidentified). Note this vehicle's horizontal hood louvers.

An Adler Type 3 Gd of Machine Gun Battalion 6 in France, 1940. Lettering: 2. (MG Schtz) Kp MG Btl. 6.

A small number of ambulance bodies were installed on the Adler Type 3 Gd chassis. The vehicles had the motor of the legendary "Autobahn-Adler", with 6 cylinders, 2.5 liters and 58 HP; the type number was W 61 K. Here such a Kfz 31 is seen in Glushzia in the spring of 1942, at the entrance to the infamous Erika-Schneise on the Volkov. The back of the sign reads, "If you are coming from here, think of Götz von Berlichingen."

Between 1932 and 1934, Daimler-Benz built 147 machine-gun cars on the Adler Standard 6 chassis. The vehicles had 8 mm armor in front of the grille and around the fighting compartment. The motor hood, though, was unarmored on the sides, which appears to have sealed the fate of this vehicle of the 1st Cavalry Division before La Rochelle in June of 1940.

In June of 1932 the merger of the Audi, DHW, Horch and Wanderer created the Auto Union KG. Horch and Wanderer immediately began to build the military Kübelwagen. The version produced from 1934 to 1937 on the Horch 830 R "with the chassis of the medium Pkw (o)" became very widespread, with over 4000 built. The motor was a powerful 3-liter Horch V8 producing 62 HP (later 3.2 liters, 70 HP). This is a platoon leader's car of the 3./Engineer Battalion 18, with lettering on the left fender. Leutnant Oberempt fell at Misikoff on 9/14/1939.

A Horch 830 R of the SS on the march into Austria on March 12, 1939. Original text: "Salzburg greets its German brethren jubilantly."

"Medium off-road Pkw with towing apparatus (Kfz 12) with the chassis of the medium Pkw (o)" of the 14th Company, I. R. 61 (WH-72 301) with 3.7 cm Pak on a training march. This Horch 830 R has shortened rear bodywork to make room for two containers of ammunition and equipment.

A Horch 830 R of the 2nd Platoon of Engineer Battalion 14 shortly before the war began. At left is the platoon leader, later Oberstleutnant Friedrich Masch. At right is a motorcycle messenger with a 750 cc BMW R 12.

This picture of a wrecked Horch 830 R comes from an album of the "Spielhahnjäger." The poor training of the driver was often to blame for such accidents. According to HDv 130/1 E (3/16/1942), the driving school course for wheeled vehicles lasted only two weeks, during which "for reasons of fuel conservation, a maximum distance of 150 kilometers must not be exceeded"!

General Konrad (49th Mountain AK) in a Horch 830 R with interesting camouflage paint. The picture was taken on 9/4/1943, but the location is not known.

More than 2800 of the Wanderer W 11 (6 cylinders, 2.5 liters, 50 HP motor) made by Auto Union from 1933 to 1936 were delivered. It was used mainly to tow the light Pak gun. This picture shows a vehicle of Panzerjäger Unit 31 in the ruins of Kukov, Poland, in 1939.

More than 2700 of the Wanderer W 11 (3.0 liters) were built from 1937 to 1941. It differed from its forerunner in having a larger (3 liter, 60 HP) motor, and is recognizable by the round front fenders and side doors.

A 3-liter Wanderer W 11 of the Engineer Battalion of Bicycle Battalion 402 in Estonia in August of 1941.

While the 3.7 cm Pak gun is moved across the Ganja River (Estonia) on a wobbly pontoon bridge by manpower, the 3-liter Wanderer W 11 proves that it has 55-cm wading ability.

The Wanderer W 23 S was modified from the standard W 23 and had the same motor (6 cylinders, 2.7 liters, 62 HP), but had a rigid rear axle instead of the usual swing axle.

A Wanderer W 23 S next to a Panzer IV with "Stummelkanone", followed by an Opel Blitz 3-ton Type S of Panzer Group 2.

A technical stop for the "Medium Survey Car (Kfz 16)" and "Medium Alert Car (Kfz 16/1) with Medium Pkw (o) Chassis" Wanderer W 23 S of Observation Battery (Pz.) 102/9th P.D. near Belaya Zerkov in July of 1941. At the far left is the anti-aircraft car (Kfz 4), one of which was intended for each sound- and light-measuring unit and staff battery.

The Russian mud soon brought the off-road capability of the Wanderer W 23 S to an end, as it did for this car of the Reich Work Service.

A whole series of requisitioned convertibles were rebuilt in makeshift fashion as Kübelwagen, like this Wanderer W 23 of the 31st I.D., identified as a battalion staff with platoon leader's car, near Brobuisk. At right is a 3-ton Borgward truck.

The BMW Kübelwagen were very rarely used by the Wehrmacht. This is a BMW 309 serving as a survey car of Engineer Battalion 14 in Poland. The performance of the two-seater with its 4-cylinder, 0.9 liter, 23 HP motor, built in 1934-35, was insufficient for field service.

The best-looking of all bucket-seat cars was probably the Mercedes-Benz Stuttgart 260 made by Daimler-Benz. About 1500 of them were built from 1929 to 1935, and had a 6-cylinder, 2.6 liter, 50 hP motor. The picture shows the Anti-Aircraft Training Company of Hauptmann von der Mosel on parade in Berlin in 1936.

In the notorious "Dead Man" and "Hill 304" battleground area of World War I, an advance unit of the 3./Panzerjäger Unit 36 had the task of taking possession of the bridge over the Forges Brook near Bethincourt on 6/13/1940. But the French engineers blew up the bridge so hastily that they and their truck remained on the German side and were captured. A short time later, precisely aimed French artillery fire began. The truck was first to take a direct hit, which killed most of the prisoners. All three limbers and the platoon leader's car were burned out; the platoon leader and five men were killed.

On 9/14/1940 Unteroffizier Hansludwig Huber (left) and J. Merkel visited the battle site. This picture shows the Mercedes-Benz Stuttgart 260 of the fallen platoon leader, Reserve Leutnant Weber.

The Mercedes-Benz Type 170 VK (V = front engine, K = bucket seats) was, after the VW Kübelwagen, the Wehrmacht's most often-built bucket-seat car, and was very popular, particularly for its reliability. The car (4 x 2) had a four-cylinder, 1.7 liter, 38 HP motor and usually reached the troops as the three-door version, Kfz 2 and Kfz 2/40.

This "Intelligence Car (Kfz 2) with Chassis of the Medium Pkw (o)" of the 101st Jäger Division sank in mud up to its axles in a mudhole near Kharkov in the autumn of 1941 and had to be towed out. Even a four-wheel drive uniform Pkw would have been stuck here!

The 10th Panzer Division advances toward Viasma in the autumn of 1941. Storekeeper Gerhard Kesenhagen's small repair vehicle (Kfz 2/40) rides amid medium light towing vehicles of the 2nd Battalion of Panzergrenadier Regiment 69. "Our Mercedes-Benz Type 170 VK never let us down – but it had to be abandoned with frozen brakes in rear-guard action before Moscow."

This Mercedes-Benz Type 170 VK belonged to the I./JG 52 and is seen here at a refueling stop in Appeldoorn on its way from Soesterberg to Utrecht on 6/18/1941. A three-door vehicle, its markings on the front as well as the trunk lid can be seen here. The fuel tank of this type, as usual, was in the motor compartment, so that a fuel pump could be eliminated. This picture appears to confirm the opinions of experts that before the war, some Luftwaffe vehicles were spray-painted with RAL 7019 gloss.

Mercedes-Benz Type 170 VK as a radio car (Kfz 2) in Romania. The antenna is fitted with a cross to limit the radio range when close to the enemy.

This three-door Mercedes-Benz Type 170 VK of Mountain Käger Regiment 85 was taken by rail to the "Leningrad encirclement front" (autumn 1942).

Radio practice by an unidentified unit in France, 1943-44. The picture shows a Mercedes-Benz Type 170 VK as a radio car with Staff Antenna E (or V) for the fairly rare Radio Listening Receiver E (or V), which was used by short-range radio reconnaissance troops.

A Kfz 2 Mercedes-Benz Type 170 VK (probably of Bicycle Battalion 402) at the Volkov in the winter of 1941-42. At right is a captured Russian BA-10 lettered "Gneisenau" on the door, with German crosses and air recognition markings, plus swastikas on the headlight covers.

This Mercedes-Benz 170 VK of the "Naval Commander, Channel Coast (MBK)" in Le Havre features the vertical rear, which was apparently customary for the rare four-door version. A night-marching device was obviously considered superfluous in the navy.

Shitomir, 1942: A Mercedes-Benz Type 170 VK with nonstandard bodywork (running boards!). In the background is a light field wagon (Hf1) which remained an essential Wehrmacht transport vehicle despite all motorization.

Other than the heavy uniform Pkw, the vehicle shown here was the only bucket-seat car built by the Ford works. It was a test model with the chassis of the American one-ton Ford V-8 of 1939, with four-wheel drive and V8 motor. The body was built by the Papler firm of Cologne. A series of 1500 units was apparently planned, but never built.

Condor Legion, Spain, 1937: The telephone squad of the 3rd Company, Aerial Intelligence Unit (mot.) 88, near Bilbao. The squad leader's car with a field cable-laying trailer, a two-seat "Intelligence Vehicle (Kfz 2) with Chassis of the Light Pkw (o)" of the Hanomag 4/20 type. In the background is a Krupp L 3 H 163. The LC symbol for Legion Condor has not been confirmed in any previous official document. The number LC-28 673 appears to have been painted by hand.

The Hanomag 4/20 (or 4/23 with a proud 23 HP) was built in 1930 and was fully out of its depth when off the road. On the other hand, one could always carry the light car around as the occasion demanded and get it to its goal, an advantage that probably inspired the troops to make rude remarks!

The bucket-seat version of the Phänomen Granit 25 H was built in very small numbers, as the very useful chassis was fitted almost exclusively with ambulance bodies (Kfz 31). The 4-cylinder motor was air-cooled, displaced 2.5 liters and produced about 37 HP.

The II./Mountain Artillery Regiment 79 (1st Mountain Division) prepares to cross the Maas in May of 1940. In the foreground is a Phänomen Granit 25 H in bucket-seat form with an unfortunately typical variety of tire sizes and types. At right is a 7.5 cm Mountain Cannon 15, at left an "Ammunition Carrier on Panzerkampfwagen I (Sd Kfz 111)", only 51 of which existed.

The 2nd Mountain Division was developed from the 6th Division of the Austrian Army and thus had a number of Steyr Type 250 Kübelwagen, a reliable five-seater with water-cooled 4-cylinder opposed motor (1.2 liters, 25 HP). 1200 of them were built. The picture is labeled: "With our leader's car at breakfast in the Oesterdal (Norway) on 9/30/1942."

While some 19,000 of the "Steyr 1.5 Ton Truck, Model 1500 A" with the 8-seat personnel car body were produced, the same chassis with cabriolet body (the so-called Commander's Car) reached the troops only in small numbers. The all-wheel drive vehicle had an air-cooled V8 motor (3.5 liters, 85 HP) which won much praise for its reliability.

This picture shows a Steyr command car of the Waffen-SS. The number SS-02047 could indicate a test or delivery drive.

Only about 600 of the Stoewer Type M 12 (4 x 2) of 1935-36 were built. It had an eight-cylinder, 3-liter, 60 HP motor. The picture shows a radio car (Kfz 15) with the b 1 or f canister radio set.

A Stoewer M 12 as an intelligence car (Kfz 15) of the 1st Company, Mountain Intelligence Unit 54 in France in 1940.

For lack of vehicles, even the Opel P 4 (with reinforced chassis) was fitted with a bucket-seat body. The vehicle had a 4-cylinder in-line, 1.3-liter, 26 HP motor and was surely a rarity among front-line units.

The "Light Personnel Car 1.3 Tatra Series 57 K (D 662/15 of 10/29/42)", of which 6000 were presumably built from 1941 to 1943, is, oddly enough, very rarely seen in photographs. The car had an air-cooled 4-cylinder opposed motor, 1.3 liters, 23 HP. This picture shows a Kübelwagen of the SS Panzer Engineer Battalion 2 "Das Reich" during the occupation of France in November 1942 (Operation "Attila"). It carries the tactical symbol of the leader of an engineer company (2nd Company).

The Volkswagen Kübelwagen (4 x 2) was the most widespread and surely one of the most popular bucket-seat cars of the Wehrmacht. It was light, easy to maintain, and had a reliable 4-cylinder opposed motor, originally (to 1943) 1.0 liter, 23.5 HP, later 1.1 liters, 25 HP. According to Volkswagen factory records, the following numbers of Type 82 vehicles had been produced as of 4/10/1945:

37,320 "light Pkw"
3326 radio cars
7545 intelligence cars
2324 repair cars
273 survey cars

The many detail changes during series production, which interested collectors and fans have discussed fervently for years, cannot be dealt with here.

The picture shows a "light Pkw K 1, Type 82" of the Luftwaffe, jacked up onto two uniform canisters for inspection. In the background is a VW Schwimmwagen. The Stabsgefreite wears the vehicle maintenance emblem (silver?) on his left forearm. The picture was taken in Hungary in the autumn of 1944.

Oberleutnant Paul Felder with a VW Kübelwagen in front of a pitch-black Do 217 M (DB 603 in-line motors). In the background is a "Night Disturber" Type 66 Arado. As of September 1943, Felder was Captain of the 4th Squadron of Night Reconnaissance Unit (F) 1. He was awarded the Knight's Cross on 2/29/2942. The squadron saw action in the central and northern sectors of the eastern front (for example, night reconnaissance at Leningrad) at the beginning of 1944.

This VW Kübel Type 82 is not driving through North Africa but, according to the text, "on the road to Stalingrad." All the occupants wear safety goggles – except the driver. This was a vehicle of Field Police Unit 521 (mot.), much of which escaped from the Stalingrad pocket.

Retreat of the "Grossdeutschland" Panzer Corps to the Dniester in the spring of 1944. A VW Type 82 Kübelwagen sits in the mud amid an interesting variety of vehicles.

A VW Kübelwagen Type 82 of JG 3.

In position near Kerch in the spring of 1944. In the background is a 7.5 cm Assault Gun 40, Type G, in front a VW Kübelwagen Type 82 of Flak Unit 257, which suffered heavy losses in Crimea.

VW Kübelwagen Type 82 of JG 53. In the background is a captured Fiat Type 626 truck, in front of a likewise captured Savoia-Marchetti SM 82 (T.G. 4?). This VW of a later series lacks the "ramming protection" on the rear and the covered louvers for inside ventilation).

Oberzahlmeister Pfaff of the 1st Unit, Panzer Regiment 11 in Hungary in 1944. A VW Kübelwagen Type 82 with "ramming protectors" on the rear.

The VW Kübelwagen was planned for a wide variety of tasks. Here is a "Starter Drive Type 198 for Kfz Motors on Light Pkw K 1, Type 82", with which not only, as here, the motor of an Opel Blitz, but also that of a "Panther" tank could be started. The starter drive, plus its shaft and connections, was stored on the floor of the VW. To attach the bracket, the installation of, among other things, an additional transverse member at the rear was required.

According to D 635/19 of 12/8/1944, spare parts were to be ordered from the "Volkswagen GmbH, City of the KdF-Wagen, Customer Service and Spare Parts Department."

In case of a missing or damaged front axle, the VW Kübelwagen could be brought to the nearest freight yard easily and elegantly by sled. A vehicle of the 3rd SS Panzer Division "Totenkopf" is seen in the Staraya region of Russia.

The VW Schwimmwagen (4 x 4), "Light Passenger Vehicle K 2s Type 166" (according to D 662/13 (12/5/42)) was produced from 1942 to 1944, with 14,500 built. It was meant to replace the motorcycle with sidecar in the SS divisions and thus was originally given the designation of "Kradschützenwagen." Little by little, chosen units of the Army and Luftwaffe (engineers, paratroops, etc.) were equipped with the Schwimmwagen. The vehicle was extremely popular among the troops, especially for its four-wheel drive and resulting off-road capability and its multiple uses. Its amphibian capability was put to the test only rarely; and in entering and leaving bodies of water in strong currents or waves, what it could do was very limited. The picture shows a "Type 166" of Armored Engineer Battalion 37 during training on a "duck pond" in Brittany in the spring of 1943.

Montenegro, 1943. This VW Schwimmwagen, Type 166, of an unidentified mountain Jäger unit (probably of the 1st Mountain Division) clearly shows its three-bladed propeller, which could be folded down on a hinge (visible over the exhaust). When it was folded down, the end of the driveshaft projecting from the rear would fit directly into the matching part of the intermediate section of the propeller drive. The speed could be regulated by the gas pedal; the top speed in the water was about 10 kph, and only forward motion was possible.

A VW Schwimmwagen, Type 166, of Armored Intelligence Unit 27 in Romania in the spring of 1944. Note the grille on the air exhaust opening in the rear wall and the two fillers for the divided fuel tank (2 x 50 liters) beside the spare wheel. The central filler was for the oil tank of the central lubrication system. Oddly, the Type 166 had no standard night-marching device.

The Waffen-SS showed a keen interest in the VW Schwimmwagen from the start and had a number of interesting test models built. This picture shows an amphibian machine-gun car, a hitherto unknown blend of Type 128 ("long" Schwimmwagen), of which only about thirty were built, and basic elements of Type 166. In front of the passenger seat, the front body is cut out and carries an MG 34 on a fixed turning mount; the front windshield is thus made smaller. Despite the warlike appearance, the crew had no protection from enemy fire, not even that of infantry weapons – this was surely one reason why this vehicle never reached the troops.

Three-Axle Cars

In the 1920s, the opinion prevailed among German and foreign military technicians that off-road capability was attained most easily with a three-axle chassis design. Despite the great technical efforts, the three-axle vehicles never came near their intended performance when off the road, and were soon given up by most manufacturers. The spacious Mercedes-Benz Type G 4 (6 x 4) would not have stayed in production if the high party officials, particularly Adolf Hitler himself, had not taken a special interest in this exclusive and attention-getting vehicle. From 1933 to 1939, 72 of them were built with various motors (8 cylinders, 5.0 liters, 100 HP; 5.3 liters, 110 HP) and bodies (some with armor). The allowable gross weight was 4.5 tons, the fuel consumption 28 to 38 liters per 100 km.

The picture shows a G 4 (probably at the Führer's Wolfsschanze headquarters) with the flag of the "Führer and Supreme Commander of the Wehrmacht."

A Mercedes-Benz Type G 4 with escort troops, photographed during Hitler's visit to I.R. 30 in Poland in 1939.

Krupp also built a three-axle command car on the chassis of the light L 2 H 143 (Krupp-Protze) truck. It was displayed at the Vienna Spring Fair in 1940.

This slide shows a Krupp command car in southern Russia, with General Paulus sitting in back. The vehicle had a sloping motor hood, similar to that of the "Protze", and the unicorn symbol of the 6th Army on the rear.

The Praga firm also produced a three-axle off-road car, the Type AV, from 1936 to 1939. It had a six-cylinder, 3.5-liter, 80 HP motor. This picture shows a vehicle of the NSKK in Poland in 1939.

Praga Type AV of Engineer Battalion 10 near Warsaw in 1939. At right is a 1938 Chevrolet with WH-131313 registration.

Skoda Superb 903, a command car with a 6-cylinder, 3-liter, 73 HP motor, small numbers of which were used by the Wehrmacht.

There was also a "Command Car" Skoda Type 640, though this picture shows a vehicle with the body of the personnel carrier, used by Engineer Battalion 11 in Romania.

Requisitioned German Civilian Cars

The multitude of models used by the Wehrmacht was the nightmare of all storekeepers. In the hinterlands in particular, passenger cars of every imaginable make and model were used. Obtaining spare parts presented insoluble problems, and the lack of a cheap part (such as the distributor rotor) often led to the total loss of a vehicle. As of 12/1/1938, passenger vehicles made by the following major manufacturers were licensed:

Adler: 90,087, Auto-Union: 305,514 (Audi: 1306, DKW: 233,704, Horch: 13,057, Wanderer: 57,447), BMW: 43,085, Borgward: 33,946, Daimler-Benz: 122,727, Hanomag: 44,391, Opel: 480,464, Steyr: 24,807, Stoewer: 6528, Fiat: 43,084, Ford: 101,837, Others: 9,139.

A typical advertisement from prewar times.

The little Adler Trumpf Junior Cabriolet (built from 1936), with its four-cylinder, 1-liter, 25 HP motor, also had to do military service. Whether the car, seen here with the 5th SS Panzer Division "Viking", was at all suitable for field service was of no importance.

A classy car of the 1930s was the so-called "Autobahn Adler", whose streamlined body was quite in the spirit of the times. This is an Adler 2.5 liter sedan, built 1937-1940, 6 cylinders, 58 HP, Ambi-Budd body, with Cycle Battalion 402 in France.

The rear view of an Adler 2.5 liter with interesting markings.

According to a directive of 10/31/1935, all service vehicles used by administrative officials had to carry the "Reich Service Flag", as does this Adler Diplomat Pullman Limousine with Berlin registration.

The 1st Mountain Division used this ex-Czechoslovakian Aero 30 of 1934 (2 cylinders, 1.0 liter, 28 HP). The registration is not recognizable. The tactical symbol with added WuG (Weapons and Equipment) can be seen. In the background is a Hansa-Lloyd truck. The picture was taken at Königsfeld.

The Auto-Union was founded in 1932 as a merger of the four named firms – some of which were having major difficulties. The "Volksgenossen" who were inspired to place an order at that show had only short-term pleasure with their cars before they were requisitioned, denied a permit or, in lucky cases, "given a chevron."

The Audi 3.2 liter Type 920 (6 cylinders, 75 HP) was first built in 1938-39 and thus was rarely seen. This picture shows a convertible with a general's pennant in April 1941 form. Unfortunately, the photographer had his eye on the charming intelligence assistants more than on the Audi or the interesting cars in the background.

Despite widespread prejudice, two-stroke motors were apparently requisitioned too! This picture sows three DKW Reichsklasse (2 cylinders, 0.6 liter, 18 HP) or Meisterklasse (2 cylinders, 0.7 liter, 20 HP). The car at left has the top-mounted windshield wipers of the later model. The cylindrical fuel tank in the engine compartment can be seen, with the shift rod passing through. Markings: Ambulance Platoon 2. Registration: Province of Westphalia.

DKW Reichsklasse of the Reich Work Service, with masked headlights and RAD license number. This low-powered front-drive car with its wooden body covered with imitation leather was really not intended for such conditions!

The DKW Reichsklasse weighed only 750 kilograms and could easily be jacked up for a tire change. This vehicle does not seem to have been given any camouflage paint (decoration!) but had a masked headlight. At right is an Oberstfeldmeister of the RAD.

The DKW Front Luxus Sport (an open two-seater) could reach 90 kph with its two-stroke, two-cylinder motor (0.7 liter, 20 HP), and thus was best suited for use by the medical corps – though it must be feared that this little racer would meet a quick end in the field!

Requisitioned cars made by Horch were found frequently in the field units because of their powerful motors. This is a Horch 830 BL loudspeaker car of Propaganda Company 621 (note the "eye" on the right fender, 18th Army, northern Russia) in its Pullman limousine form. The car was built with V8 motors of 3.5 liters with 82 HP, or 3.8 liters with 92 HP.

The Horch 830 BL in Pullman Limousine form, used by the commanding general of the 101st Jäger Division near Kharkov in the autumn of 1941. At left is a Mercedes-Benz Type 170 V.

This Horch 830, rebuilt as a light truck, bears the tactical symbol of a motorcycle rifle company.

A beautiful Horch 853 convertible with 8-cylinder, 5-liter, 120 HP motor, built from 1937 to 1939 and even then an unattainable rarity. Unfortunately, no data is available, but in view of the covered lights and the seven-digit license number (WH-1 076 699), the picture seems to come from the latter half of the war.

"White lettering on a black background" was surely only a temporary situation until the issuing of an official license, as seen on the motorcycle at the right rear. What may have become of the splendid BMW R 16 with the jazzy sidecar, or the Horch 780 convertible (8 cylinders, 5 liters, 100 HP) in the background?

Generalleutnant Friedrich Carl Cranz, the Commander of the 18th Silesian I.D., used a Horch 8 Type 500 B Pullman Cabriolet (8 cylinders, 5 liters, 100 HP) during the French campaign. It lacked camouflage paint and covered lights, but its markings were exemplary. From left: General's pennant, black "WH" in a white oval, division emblem (black-white-red), "IK" civilian registration for Lower and Upper Silesia, division pennant (black-white-red), commander's chevron, and a white wreath (for Cranz) as a personal symbol.

Rear view of Generalleutnant Cranz's Horch on the road toward Amiens. Cranz died on March 24, 1941 after a shooting accident at the Neuhammer training center.

The Wanderer works also belonged to Auto-Union. This picture shows a Wanderer W 23 sedan of the staff of Engineer Battalion 212 on the border of Luxembourg in 1940. The registration number "III A" indicates Stuttgart, the pennant means "battalion." The storekeeper sergeant standing beside the car is Michael Frank, my father, who has been missing since June 1944 near Minsk.

The Wanderer W 23 convertible of a Luftwaffe unit (WL-159002) at Mont-Saint-Michel in France. The W 23 had a 6-cylinder, 2.7-liter, 62 HP motor and was built from 1937 to 1941.

Wanderer W 50 (2.3 liters) four-window convertible (body by Gläser) with the tactical symbol for "staff of an artillery unit" in Greece. This model was built from 1936 to 1938 and had a 6-cylinder, 2.3=liter, 50 HP motor.

A Wanderer six-window sedan (probably Type W 250, built in 1935) of the motor vehicle unit at Wilhelmshaven, at a civilian "Standard" gas station.

The Bayerische Motoren Werke – BMW – did not produce any vehicles for the Wehrmacht during the war (except a few light uniform Pkw). The troops obtained exclusively requisitioned civilian BMW cars: they were beautiful, elegant vehicles, but they were not built for rough road conditions.

The picture shows a BMW 326 two-door convertible (6 cylinders, 2 liters, 50 HP) of Ambulance Platoon 54 on the way to the Duka Pass in 1939. At right is one of its frequently used stablemates: a BMW R 61 (600 cc) with Steib sidecar. All the vehicles still bear their original paint, but have WH license numbers.

The same BMW 326 from the left rear. The "S" on the tactical symbol, for "Sanitätsdienst" (Medical Corps) is easy to see.

The BMW 326 also seems to please these Belgian soldiers. The car, from the province of Westphalia, seems to have been oversprayed with a glossy camouflage color; the "P" on the tactical symbol may stand for "Motor Pool."

A BMW sedan of the III./JG 3 while being transferred in June of 1941. It may be a BMW 326; the two-liter models differed only in minor details and cannot be identified easily.

"Swimming lessons in the field," demonstrated with a BMW 320 convertible of the 1st Mountain Division.

On June 30, 1941, an advance guard formed of various fast units and commanded by Oberst Lasch, reached the city of Riga. This picture shows the market place, near the water tower, still smoking, with two assault guns (probably of Assault Gun Unit 185) and one self-propelled anti-aircraft gun securing the area. In front is a BMW convertible, plus an unknown car (Ford?) with "Pol-46885" plates, indicating the zone of the 1st I.D., as well as an illegible runic symbol. At left is a car that much resembles a 1940 Pontiac; the "R" could stand for the Reinhardt Panzer Corps.

Josef "Pips" Priller, one of Germany's most successful fighter pilots, also drove a first-class BMW. Here he is seen as Commander of the II./JG 26, with a BMW 327 (or 327/28) Sport Cabriolet, in front of his Fw 190. The vehicle shows no license number – could it have been unregistered and illicit?

Necessity is the mother of invention: With the enemy at one's heels, one could load a large amount of luggage onto a BMW 327 (or 327/28) Sport Cabriolet (WH-887990) and retreat.

This Panzer man of the 6th Panzer Division stands proudly beside a BMW 319/1 sports car (6 cylinders, 1.9 liters, 55 HP). In civilian life, such a car probably would have been an impossible dream for him!

A Daimler-Benz advertisement of 1942.

Daimler-Benz AG: Mercedes-Benz Type 170 V Cabriolet B with JG 3 in Russia in 1943. The vehicle had a 4-cylinder, 1.7 liter, 38 HP motor and was used in great numbers.

Ready to attack Riga, June 1941: Soldiers of Bicycle Battalion 402, a Mercedes-Benz Type 170 V four-door Cabrio-Limousine and a Mercedes-Benz Type 170 VK radio car, a heavy uniform Pkw and a Wanderer Kübelwagen; at left are supply-train units with steel field wagons.

Well marked with squadron and group symbols left and right, plus the tactical symbol of a light fighter squadron, this Mercedes-Benz Type 170 V convertible of the III./JG 52 stands in front of a Focke-Wulf "Weihe."

Athens, May 1942. The photographer certainly just wanted to show the traffic policeman and had no idea what a rarity he was capturing on film: a Mercedes-Benz Type 170 H Cabrio-Limousine. One of the few Mercedes cars built with what was then a very modern streamlined body and a rear engine. Since the 170 H was not very satisfactory in either its handling or its roominess, and also cost 600 Reichsmark more than the 170 V, very few of them were built. By its markings and date, it belonged to the 1st Mountain Division.

"Eye to eye" with a French 32-ton tank is this Mercedes-Benz Type 230 Cabriolet B. Fortunately, the Char B1 was abandoned by its crew near Mulhouse!

A Mercedes-Benz Type 230 four-door sedan with roof hatch, of Propaganda Company 649, south of Mostar on May 5, 1941.

Mercedes-Benz Type 230 convertible with civilian registration for the Palatinate. The photo comes from an album, "Comm.AOK Engineers AOK 2." The emblem has not been identified to date; the sidecar at the rear shows the emblem of the 17th I.D.

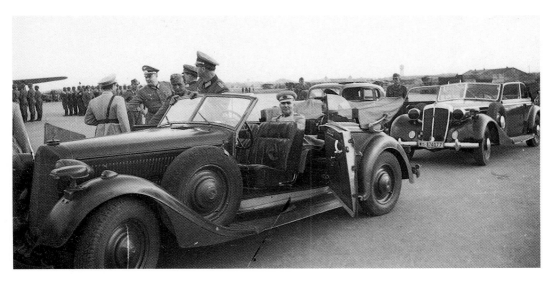

"Grand Central Station" for Germany's allies: A Mercedes-Benz Type 320 Cabriolet B with high Romanian officers, behind it a Horch 930 V four-window cabriolet (WH-874077).

A Mercedes-Benz Type 320 in a rare version, a four-door convertible, at the Maas in June 1940.

The granting of vehicles to high-ranking troop leaders was undertaken by Hitler personally. The higher the rank, the more luxurious and powerful the service car had to be. Generaloberst Fedor von Bock, as leader of the Army Group North in Poland, received this wonderful Mercedes-Benz Type 540 K. The car had an eight-cylinder, 5.4 liter, 115 HP motor, which turned out a proud 180 HP when supercharged, but with its long wheelbase and overall weight of 2.6 tons was quite helpless off the road. Here a little VW Kübel surely would have come through without problems.

Hitler's first visit to Karlsruhe in 1938. Center: Mercedes-Benz Type 540 K (probably an armored version), left and right also 540 K cars with different bodies – in this luxury class, the bodies were almost always built to suit the individual. In the left background is Heinrich Himmler in a 5-liter Horch with convertible body.

This Mercedes-Benz Type 200 four-door sedan, built in 1935-36, was used by the III/JG3.

In contrast to Hitler's four-ton Mercedes, here is the "smallest passenger car of the Wehrmacht": a DIXI 3/15 built in 1927, with a four- cylinder, 0.7-lite, 15 HP motor and a gross weight of 700 kilograms. – Despite the wartime headlights, the grinning soldiers of Engineer Battalion 6 do not seem to be taking the car completely seriously!

Ford cars were used in great numbers by the Wehrmacht. This is a Ford V-8 four-door sedan, built 1935-36, Model 48, of the 168th I.D., followed by an older Opel Blitz. The corduroy road and drainage ditch were built just right – presumably by the RAD.

A Ford V-8 convertible, Model 48, a five-seater of 1935-36. The car had an eight-cylinder, 3.6-liter, 90 HP motor.

Rolling southward, advancing on Kiev. "South of the Pripyet Marshes, the roads are sandy and the vehicles stagger forward in deep ruts. When the guns are taken care of, the towing tractors are used to tow cars," writes the technical Kriegsverwaltungsrat Hansludwig Huber about this picture, taken in August 1941. A Ford V-8 "Standard" sedan, 1937-38, two-door sedan, Model 48 G 81, with the white mountain boot of the 99th Light I.D. (later 7th Mountain Division) and registration for Upper Franconia, followed by a light uniform Pkw.

Ford V-8 "Spezial", 1938-39 five-seat convertible, Model 48 G 81, of an intelligence unit, with Berlin plates.

A war correspondent wearing hobnailed boots stands on the roof of a Ford limousine with his camera equipment as he films the burning port of Dunkirk. The Ford V-8 "Special," four-door, 1938/39, Model 48, G 81, is equipped with running boards and hand grips for easier access.

There were a number of conversions of the Ford V-8, for example panel boxes, frame antennas and, as seen here, a towing vehicle for the 20mm Flak 38. These Ford V-8 "Specials," 1938/39 were converted by Papler of Cologne.

Staff officers of the 18th Infantry Division during a rest on the "East-West Rollbahn." The expression "Rollbahn" was obviously not invented in Russia. On the right a Standard Medium Car, left a 1937-39 Ford V-8 two-door convertible.

Clearly visible in this photograph are the projecting rear end and the unpractical rear fenders of the 1937-39 Ford V-8 two-door convertible. The car belonged to JG 52.

The prototype of a Ford V-8 "Spezial", a 1938-39 four-door sedan, Model 38, with an Imbert wood-gas generator at the rear. The heater is carefully concealed, with two air scoops installed beside the rear windows and the "wood tank" covered on the roof and reachable by steps; the spare wheel has been moved from the rear to the front fender. Series production did not take place. In this picture, the "red angle" on the license plate is clearly seen.

Ford-Eifel also made this test model with wood-gas drive. The picture shows a two-door sedan built in 1938, Model 20 C, with the gas generator at the rear, the duct carrying the gas to the motor, and a gas filter in front of the radiator.

The Ford Eifel was a common vehicle with a four-cylinder, 1.2-liter, 34 HP motor. The picture shows a 1937 convertible sedan, Model 20 C (with Stuttgart plates) of the Hermann Hoth Panzer Group near Smolensk in the summer of 1941. On the other side of the road are Panzerjäger vehicles of "4.7 cm Pak (Czech) on Pz Kpfw Renault R 35 (French)" type with the triangle symbol of the 4th Army.

Ford Eifel Model 20 C two-door sedan, built in 1937, in one of the numerous "national reliability trials", here in East Prussia in 1939. The original text of the Ford factory journal says. "Only the best of the best can survive this 1000-kilometer, three-day test, as well as the combat service of men and the performance of motors and ruggedness of materials." These events, mostly organized by the NSKK, were meant to provide information on the suitability of the various models for military use.

The Ford Taunus was the German Ford factory's first independent development, and was introduced only a few months before the war began. Photos of this car, as of the KdF Volkswagen, are thus very rare. At Taunus, the motor and gearbox of the Eifel were used and the completely new chassis and body developed according to modern concepts. The pictures came from Unteroffizier Fröba, who was less renowned for this car than for his crash-landing on the horse-race track in Karlstadt, Sweden on October 24, 1940. His Bf 109 E, number 0820, "white 3", of 4./JG 77 is presumably still in Sweden today.

Largely forgotten today are the passenger vehicles made by Hanomag, solid, handsome cars, most of which were built only in small series. Here is a Hanomag Rekord convertible, obviously the platoon leader's car of a bridge platoon. The Rekord had a four-cylinder, 1.5-liter, 32 (35) HP motor and was built from 1934 to 1938; it was also offered with a 1.9-liter Diesel engine, one of the first cars in the world to have one.

Hanomag Rekord two-door sedan in the ruins of Amiens in 1940.

The Hanomag Sturm had a straight six-cylinder, 2.3-liter, 50 (55) HP motor and was the largest car that Hanomag built. This picture shows a car of the Engineer Battalion of SS Division "Das Reich" near Rshev early in 1942. Once can see how hastily the car's occupants had to move.

A Hanomag Sturm four-window convertible (body probably by Hebmüller) of the 18th I.D. on its way back from France.

The Hanomag Sturm convertible, this one probably with body by Gläser, was exceptionally rare. The picture shows a vehicle of Bicycle Battalion 402 near Maintenon (fighting against French colonials) on 6/16/1940.

Another rarity was the Hansa 1100, a product of the "Hansa-Lloyd und Goliath-Werke Borgward und Tecklenborg." The car (here with Bicycle Battalion 402 in France) had a 4-cylinder, 1.1 liter, 28 HP motor.

The Maybach SW 38, an exclusive limousine with a weight of 2.7 tons and a 6-cylinder, 3.8-liter, 140 HP motor that allowed a top speed of 140 kph. Unfortunately, it is not known who allowed the use of WM-5945 in the "homeland war zone" and thus consumed great quantities of gasoline (20 to 30 liters per 100 km) badly needed at the front. Such high-class cars often survived the war and found enthusiastic buyers, particularly in America.

Advertisement for the Opel Blitz and Kapitän of 1943.

Opel cars were often seen in the Wehrmacht because of their high percentage of the market. This is a 1936-37 Opel P4 two-door Cabrio-Limousine with 4-cylinder, 1.1-liter, 23 HP motor, surrounded by motorcycle messengers.

This 1937 Opel Kadett two-door Cabrio-Limousine appears to have reached the end of its road. If the emblem on it is that of the 78th Assault Division, then the car had quite a long life, for this unit was formed only in January 1943 of the 78th Infantry Division. The long license number WH-1 254 004 also suggests a late war year.

This Opel Kadett with its small 4-cylinder, 1.1-liter, 23 HP motor was somewhat out of its depth in "Stalin asphalt"! The native products in the background were naturally more at home in such conditions.

The Opel Olympia was the world's first car to be mass-produced with a self-carrying all-steel body. The car had a 4-cylinder, 1.5-liter, 37 HP motor and was built from 1935 to 1940. This picture shows a 1938-1940 Cabrio-Limousine.

"Poland, 1939" is written on this photo, which shows a highly polished Opel Olympia sedan being towed by Luftwaffe soldiers.

Opel Super 6 four-window convertible, built 1937-38, with a Luftwaffe unit. This roomy car had a straight six-cylinder, 2.5-liter, 55 HP motor and was very popular despite its somewhat soft suspension.

France or Belgium, 1940: Rear view of a requisitioned Opel Super 6 four-door sedan of 1937-38, with the emblem of the 216th I.D. Probably a staff car of Artillery Regiment 216.

Opel Super 6 four-window convertible, built 1937-38, of the 18th I.D. on the way back from France.

A technical stop for the Opel Super 6 convertible of the battalion doctor, Dr. Breitwieser, of the staff of Engineer Battalion 18 (mot.). The picture of the richly camouflaged car was taken in the summer of 1941 and shows how empty engine compartments were in those days.

A beautifully camouflaged Opel Kapitän two-door sedan of JG 3. This car, built from 1938 to 1940, had the typical hexagonal headlights and a straight six-cylinder, 2.5-liter, 55 HP motor, the same one used in the Super 6.

One of the lovely Opel Kapitän convertibles in action with the 5th Panzer Division in France, 1940. Behind it is a Henschel Type 33 with enclosed body.

The Opel Kapitän four-door sedan of the squadron staff of ZG 26 in Hanseberg, near Königsberg in the Neumark, in the summer of 1944. At this time the squadron was using Me 410's to fight against bombers and had suffered heavy losses from escort fighters; thus the squadron had been moved out of the range of enemy fighters.

On the fender is the flag of the squadron commander, on the bumper the squadron's (old) coat of arms. Note the covered headlight installed in the center.

Field Marshal Busch at the main dressing station of the 18th I.D. (mot.) in Nagatkino in the summer of 1943. In front is Dr. Breitwieser's car, an Opel Admiral 1938-39 four-door sedan, with the tactical symbol for "staff of a military hospital unit (mot.)." General Busch's care for his wounded was described impressively by Prof. Dr. Hans Kilian in his book "Im Schatten der Siege."

Cars built by Röhr were extremely rare, for this small auto builder produced only about 3000 cars in various small series from 1927 to 1935. The picture shows a Röhr junior convertible without the "red angle" but also without Wehrmacht markings, but with masked headlights. The Junior was a license-built Tatra 30 with an air-cooled opposed 4-cylinder, 1.5-liter, 30 HP motor, and was built from 1933 to 1935.

An interesting joint advertisement by the Steyr and Skoda firms for the Vienna Fair, March 9-16, 1941.

Skoda Superb of a police unit after a traffic accident (place and date not recorded). Note that this car already had all doors hinged at the front – as is customary today for reasons of safety.

A luxury-class Skoda car in front of the district school of the NS-Frauenschaft of the Prague district; it often got into the news and was promised to both Goebbels and Goering as a private car. According to "Atlas Nasich Automobilu" of Milan Spremo, it is a Skoda Superb 4000 OHV/Type 919/1940 Model with an eight-cylinder, four-liter, 96 HP motor that burned 20 to 24 liters of gasoline per 100 kilometers, and of which fewer than ten were built. A closer look, though, suggests that it is powered by a wood-gas generator concealed under superb bodywork: big rear doors to the gas generator, vents under the rear fender and a flap in the front fender for access to the blower with the mouth of the blower duct on the rear edge of the fender. A picture of the right side shows an elegantly shaped container for the fuel supply on the front fender – instead of the spare wheel.

The vehicle has obviously been sprayed with metallic paint and has a masked headlight, but no rear light. The lettering on the tires reads "Michelin, made in Bohemia."

In April 1945, the remaining population of Guben on the Neisse was rescued by a counterthrust of hastily assembled units shortly before the attacks of the Russian conquerors. As can be seen, even a Steyr Type 55 sun-roof sedan in original two-tone paint took part in the attack in company with a "Hetzer" tank! The "Steyr Baby", as the car was also known, had a water-cooled opposed 4-cylinder, 1.2-liter, 25 HP motor and was built from 1938 to 1940.

A Steyr Type 200 two-window convertible of the Luftwaffe. The car had a four-cylinder, 1.5-liter, 35 HP motor and was built from 1936 to 1940.

A dust-covered Steyr Type 220 two-window convertible of Mountain Observation Unit 38 in Poland in 1939. At left is an Opel Kadett, in the background Mercedes, Austro-Fiat, Ford and uniform Diesel trucks. This unit was established in Military Zone XVIII in 1938, with its peacetime headquarters in Klagenfurt, and thus had numerous Austrian vehicles. The Steyr Type 220 had a straight 6-cylinder, 2.3-liter, 55 HP motor and was built from 1937 to 1941.

The 1st Company of Field Police Unit 521 (mot.) turns out in Liege, Belgium. In addition to several Mercedes, one can see a goodly number of Steyr Type 220 cars with "Pol" registration. Some motorcycles still bear Austrian army numbers.

A Steyr 220 of the 1st Company, Field Police Unit 521 (mot.) being serviced in Vilna in May 1942. On the left fender is the (rare) symbol for "order-keeping services", at right the "H" of Panzer Group 4 (Generaloberst Erich Hoepner).

Autumn 1942, Soldatskaya in the Caucasus: Major von Bonin and Oberstleutnant Mützow in a Steyr 220 convertible with the symbol of the III./JG 52 on the door.

A Tatra advertisement in "Motor und Sport", April 19, 1942.

Streamlined body, rear engine, air-cooling, swing axles and a largely self-carrying body – all united already in the 1934 "grosse Tatra", a vehicle that attracted attention everywhere, even though not all its technical advances could prevail. This picture shows a Tatra Type 77 (3rd version) of the Luftwaffe. The car had an air-cooled rear-mounted 3-liter V8 motor and was built from 1934 to 1936.

Transfer of Tatra Type 87 cars by a police unit that saw war service in Yugoslavia and Italy. Note the masked headlight mounted on a cover of the central light. The Type 87 had a 3-liter V8 rear engine that produced 72 HP.

A Tatra Type 87 of the Luftwaffe as a "test car", about which no information is available. The big V8 rear engine can be seen in the upper photo.

Photos of the VW sedan (KdF-Wagen, "Beetle") are, strangely enough, quite rare. These rarities come from Dipl. Ing. Rudolf Essler-Rziha, an official of the Reich Traffic Administration in Minsk, and were taken during the evacuation of that city in the catastrophic summer of 1944. They show the Volkswagen Type 82 E (KdF body on the high-wheeled Kübelwagen chassis, 546 built) with license DR-102 115 and DR-102 116, a Mercedes-Benz Type 500 with box body, a Mercedes-Benz L 3000 S truck and a DKW Meisterklasse (or Reichsklasse) convertible, IM-135 775, probably the private property of Administration Chief von Ludendorff. Note also the number of Russian refugees who did not want to fall into the hands of the Red Army.

Captured Cars

Taking possession of all motor vehicles in a conquered land was (and is to this day) a prerogative of the conqueror. Captured vehicles were used by all the world's warring armies, not just the Wehrmacht. The picture shows the repair shop of the Daimler-Benz agency in Munich, damaged by bombs, in the autumn of 1945, with a captured Horch 8 Type 780, a Horch 930 V and a Mercedes-Benz Type 170 V, plus a wide variety of American and other passenger vehicles, being serviced.

France

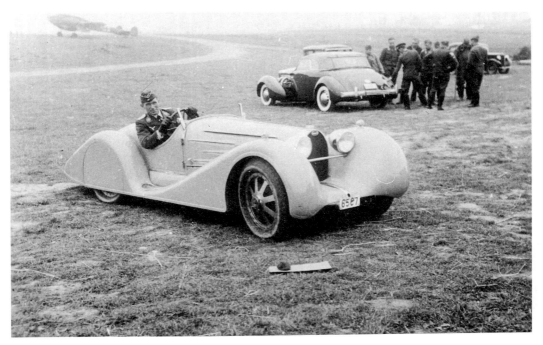

Bugatti: Oberleutnant Victor Mölders (Kapitän, 2./JG 51) in a Bugatti 35 B on Type 37 chassis, a special model with a straight 8-cylinder, 2.3-liter motor and chassis number 37,328, which can still be seen today in the Schlumpf Collection in Mulhouse!

In the background is an American Cord.

Citroen: The Citroen "Traction Avant" with its front-wheel drive, pontoon bodywork and torsion-bar suspension, was one of the most progressive cars of its time, and some 1800 of them were assembled in Germany in 1934-35. The Type 7 (4 cylinders, 1.6 liters, 35 HP) and the Type 11 (4 cylinders, 1.9 liters, 44 HP) looked very much the same. The "Traction Avant" was built from 1934 to 1957 –surely a sign of its popularity. The picture shows a captured Citroen "Traction Avant" with the emblem of the 112th I.D. and the tactical symbol for "engineer vehicle column." Note the socket of the (removed) masked headlight.

Citroen: A Citroen "Traction Avant" with the tactical symbol of an artillery unit and a non-standard license number, probably that of the field post office.

Citroen: "The good Citroen has now broken down too" is written on this photo from the 1st Mountain Division. The car bore registration number WH-567 878 and was captured near Gien.

Citroen: A civilian "Citroen "Traction Avant" just captured by the Augsburg Intelligence Unit 27. The firm's emblem (two meshing gears) has apparently been removed already.

Citroen: Naturally the Citroen "Traction Avant" was not built for such conditions! When the tire chains no longer helped, this staff car of an unidentified engineer battalion had to be towed out of the mud.

Citroen: Two Citroen "Traction Avant" and a medium uniform Pkw of the 292nd I.D. in Siedlce, Poland, in the autumn of 1940.

Citroen: Two Citroen "Traction Avant" of the Navy within sight of the Eiffel Tower. At left a Matford, at right a car similar to the 1937-38 Nash. The right Citroen has louvers instead of the usual vent flap; all cars bear the letter "V" (Versorgung?) on the fender.

A Heinkel He 111 of KG 55 prepares to take off for a night attack during the "Battle of Britain" in the autumn of 1940. Lights to illuminate the runway and the Citroen of a Luftwaffe medical unit stand ready. This "Traction Avant" also has louvers instead of flaps in the motor hood.

Delage: Even in France, the fine cars of Delage were rarities. Because of economic difficulties, the firm was taken over by Delahaye in 1935. The picture shows a right-hand drive Delage seven-seater, Type D 8-100 (or D 6-75 with a long chassis) in Luftwaffe service with the Delahaye emblem (!) on the grille.

Hotchkiss: The French auto industry also had major economic problems in the 1930s, which resulted in several closings and mergers. A typical example was the firm of Amilcar, which had been successful for years but was absorbed by the Hotchkiss firm shortly before the war. Under the name of "Amilcar compound", some 15 cars with front-wheel drive, independent suspension and self-bearing bodies were built in 1939. The first prototype had the chassis of the German Adler "Junior" and an aluminum body.

"At the Eiffel tower, August 1941" is written on the photo of this "Amilcar Compound" with registration WH-655 183, and it may possibly be the first prototype mentioned above, a photo of which has never yet appeared.

Laffly: A newly captured (normally) five-seat Laffly V 15 R VLTT communication car, of which the French Army had some sixty, and some 900 which were also built for the Wehrmacht until the end of 1940.

Matford: This firm existed only from 1934 to 1939; it grew out of a merger of the Mathis firm with the French Ford branch. The picture shows a Matford 1938 model (F 82 A or F 81 A) of the staff of the II./KG 77 in Russia in 1942.

Matford: A captured Matford 1939 Fourgonette before the city hall of Calais on 7.16.1940. The soldiers belong to JG 52 and are obviously wearing pieces of British uniforms.

Peugeot was and is one of the oldest and best-known French auto marques. This is a captured Peugeot 201, which was equipped with either cooling flaps (as here) or louvers. In the French Army this car was used as a "Voiture de Liaison."

Peugeot: Brussels, 8/31/1944: While a newsstand still advertises the "Berliner Illustrierte" and the "Adler", the German troops prepare to leave the city. At right is a Peugeot 202 with civilian plates, at left a Simca Cinq with a hand-painted Luftwaffe marking "white on black" and the (sleeve) emblem of a storekeeper sergeant.

Peugeot: The Peugeot 202 had a four-cylinder, 1.1-liter, 30 HP motor and the firm's typical headlight arrangement behind the radiator grille. The picture shows a convertible of the 6th Infantry Division in Luxembourg.

Peugeot: A Peugeot 202 withdrawing from Brussels on September 2 or 3, 1944, photographed by a Belgian civilian. The car was marked with an oversize red cross, which we hope helped!

Peugeot: This Peugeot 202 sedan of the "Leibstandarte SS Adolf Hitler" was abandoned in Louvain, Belgium in September 1944 after being hit several times.

Peugeot: The 18th I.D. (mot.) was equipped mainly with French vehicles. This picture shows a Peugeot 402 of I.R. 51 near Staraya Russia in 1942.

Peugeot: The Peugeot 402 had a 4-cylinder, 2.1-liter, 63 HP motor; some 1500 of them were used by the French Army. Here are three views of a car of the 1st Mountain Division in and around Paris in 1940. Note the double bumper, the "402" number in the central strip of the grille, and the artistic license plate.

Peugeot: The Peugeot 1.2-ton DK 5 J van was the delivery-truck version of the Peugeot 402 and was used in great numbers by both the French Army and the German Wehrmacht. This light van actually does not belong to the theme of this book, but these pictures are too good to be kept from the reader! Above: Transport column near Hammelburg, spring 1941; below: preparation to be turned over to the 7th Mountain Division in Gmünden on the Main. At right is another French truck, a Matford F 917-WS 4.5-ton model.

Renault: This firm was also one of the leaders in all kinds of vehicle production before the war. In the picture is a Renault Juvaquatre with the "green circle" emblem of the French supply-train units, followed by a Renault of about 1932 and various other captured vehicles, which the 18th I.D. caused to disappear from the harbor area at Dunkerque quickly and inconspicuously as soon as the fighting ended.

Renault: A Renault Primaquatre sedan of 1938, captured by I.R. 54 in Poperinge.

Renault: On the causeway near Tudulina in the Baltic area, this Renault, probably a Vivastella, fights its way through the autumn mud in 1941.

Renault: This Vivastella or Viva Grand Sport certainly had better driving conditions on the Champs-Elysees in Paris! The car still shows its French registration and has no headlight covers – who might be riding in it?

Renault: The market place in or near Zillebeke, with Wehrmacht vehicles, captured cars and those that will soon be in Wehrmacht hands. In the foreground is probably as Renault Vivaquatre – the various types of this marque are very hard to identify!

Simca: The Simca Cinq was a Fiat "Topolino" built under license. The photo of this captured car (4 cylinders, 0.6 liter, 13 HP) of Engineer Battalion 212 is captioned: "Aufthaler with his Mickey Mouse."

Rosengart: According to its license plate, this second-series Rosengart Supertraction 4-5 passenger coach of spring 1939 belonged to a high-ranking personage of the civilian administration in France. In the 1930s, Rosengart built small series of prestigious passenger cars on the front-drive chassis of the Adler Trumpf and, as here, the Citroen Traction Avant.

Britain

Austin: The Austin 8HP Series AP (4 cylinders, 0.9 liter, 24 HP) was a "militarized" version of this two-seater, 9500 of which were built. Captured British cars were fairly uncommon! This Austin was with the 1st Panzer Division.

Austin: A newly captured Austin 8HP of Engineer Battalion 18 in the Dunkerque area. Apparently there was room for four people in this two-seater – though without personal luggage.

Ford: An abandoned British Ford Prefect of 1938 in the Dunkerque area.

Ford: These Canadian Ford F 15's of the British Army were captured in Greece and rebuilt into "Kfz 15 (Behelf)" vehicles by Engineer Battalion 659 (mot.). In December 1941, fifteen of them were sent on their way from Athens to Russia to Army Engineer Battalion 666 (see "swallowed six" emblem). The vehicle had room for five men and a small pontoon sack in the luggage space.

Humber: Opel Blitz, Humber (perhaps Hillman) and Matford in northern France, 1940.

Wolseley Twenty Five Super Six at an unidentified naval support point. The car had a six-cylinder. 1.8-liter, 16 HP (or 3.5-liter, 25 HP) motor.

Triumph (?): The veteran British cars of the 1930s are difficult to identify. This car ran at a field air base with no markings. On account of the right-hand drive, British vehicles had their masked headlight on the right side. In the background, the motor of a Ju 88 (7A+) of Reconnaissance Group 121 is being changed.

Russia

GAZ: Cars were very rare in the Soviet Union and driven only by officials and functionaries – surely not by workers or farmers. As a result, captured Russian cars were found only rarely in the Wehrmacht. The picture shows a GAZ M1 that needs repairs. This car was built from 1936 to 1942 under license from Ford, and had a 4-cylinder, 3.3-liter, 50 HP motor. The chassis was rugged enough to be developed into the BA-20 armored scout car. Note the formed spare-wheel cover.

ZIS: Another Russian car was the ZIS 101 (or 102), a true luxury car that was produced in small numbers at the Stalin works in Moscow. The ZIS 101 was built from 1936 to 1939 and had an 8-cylinder, 5.8-liter, 90 HP motor. Original caption: "For the luxury car of the local Soviet big shot, the deconsecrated church was good enough."

USA

Buick: In the 1920s, foreign cars – especially American cars –were imported into or assembled in Germany in large numbers. By 1938, according to statistics, the number of "other cars" (made by foreign or small German firms) had sunk to only 9139. Most American cars were probably captured in France: big, comfortable cars with large-volume but thirsty motors and soft suspension, intended solely for nice smooth roads. The picture shows a 1936 Buick, which existed with 3.8 liters (93 HP) or 5.2 liters (120 HP), with five different wheelbases and many special bodies. The picture was taken by Baron von Saldern (I.R. 51), who later won the Knight's Cross, in the Dunkerque area.

Buick: By installing an Imbert generator, this gas-guzzling 1937 Buick (4.1/5.2 liters, 100/130 HP) could run on "down-home fuels" (from Kroll, "Der Gasgenerator", Berlin 1943).

Naturally one had to be photographed with "his" American luxury car. In this way, the mistaken opinion that great numbers of US cars were used by the Wehrmacht came into being, although there were really just a few individual specimens. This is the 1939 Buick of a Navy staff, at left the "Service Pennant of the Navy", at right a pennant resembling the symbol for an army battalion.

Buick of 1939 in Helsinki in 1942, with unidentified license plate, photographed by a Luftwaffe man.

Buick: General Christian Philipp as Commander of the 6th Mountain Division (yellow Edelweiss) at the Arctic front with his 1940 Buick convertible (8 cylinders, 4.1/5.2 liters, 107/141 HP).

Cadilac: This classic marque also appeared with the Wehrmacht. The picture shows a 1938 Cadillac Type 60 (8 cylinders, 5.7 liters, 135 HP; some models from that year had V-16-cylinder, 7.1-liter, 175 HP motors). The picture was taken in I.R. 54 around Dunkerque.

Cadillac-La Salle: La Salle was a branch of General Motors' Cadillac division, intended to offer high-quality cars at reasonable prices. This is a 1939 La Salle in a parade of Romanian and German officers and medalists at Odessa in 1941.

Cadillac-La Salle: A wonderful 1939 convertible with Wehrmacht license. Who knows what became of the car and the young soldier?

Chevrolet, another branch of General Motors, and one of America's most popular makes. At right is a 1937 Chevrolet (3.5 liters, 85 HP), at left a 1936 Buick, photographed in Sedan in June 1940.

Chevrolet: Highly polished and not yet camouflaged, this 1938 Chevrolet of the I./JG 77 in a splinter box. The boot symbol comes from the IV./JG 132, the tactical symbol indicates a light fighter group. Note the centrally mounted masked headlight.

Chevrolet 1938 of Ambulance Platoon 54, followed by an Austin ambulance. In "Trucks of the Wehrmacht" this Austin appears on page 154 – burned out after a fight.

Chrysler – with General Motors and Ford, one of the leading American manufacturers – was America's pioneer in building streamlined cars (Airflow). The picture shows a Chrysler Airflow of 1935 with masked headlights in Bucharest in 1941, probably the car of a member of the Romanian government.

Chrysler: This luxury car bumps laboriously across the plains of Russia; the Spiess has to get out and show the way so the car does not get stuck: good prospects for the coming snow and mud seasons! WH-308720 appears to have a new grille, so it cannot be identified definitely, but shows signs of the 1939 De Soto and 1939 Plymouth models, both made by Chrysler.

Chrysler Royal 1939 (6 cylinders, 4 liters, 108 HP) with the emblem of JG 3 "Udet." Oddly, the pennant is that of a battalion commander (infantry) of the "Hermann Goering" Division, and the emblem at right next to the Udet emblem is reminiscent of JG 1 "Oesau."

Chrysler: As this German pamphlet of March 1945 (!) shows, the 1939 Chrysler represented plutocracy in propaganda!

Cord Model 810: Only 2320 of these dream cars were made from 1935 to 1937 – understandably, for Cord was almost twice as expensive as the lowest-priced Cadillac. Production was halted after 1937, but replicas were built from 1964 to 1973. This front-drive car had a V8-cylinder, 4.7-liter motor, which produced 195 HP when supercharged.

Lille, 1940: These Luftwaffe men seem to have enjoyed their ride in a Cord thoroughly!

Dodge: Attach a few cable rolls and poles behind the fender, and it was ready: the :Intelligence Car (Kfz 15) with chassis of the medium Pkw (o):. By its displacement (3.5 liters), this Dodge was a Model DS of 1934, and naturally more of a heavy Pkw, though in the view of many a storekeeper, the boundaries were quite fluid. The car has an opening rear window as special equipment and is being ferried across the Lys on a float ferry.

Ford: An American Ford V-8 Model 68 (1936) with the "Condor Legion" in Spain. The meaning of the "CC" registration is unknown.

Dodge: In this French port (can anyone recognize it?) a whole shipload of private cars was found and unloaded by the Navy with the help of the local personnel. The vehicles were taken over by an Army unit. One of the ships bore the name "Diamant."

In the air is a 1937 Dodge; the men are working on a British Ford V-8 (1937-38), the others are unidentified.

Ford: The Lincoln and Mercury firms were branches of Ford Motors, and there were many more or less independent production facilities all over the world. This is a Lincoln Zephyr of 1936 of the 3rd Company, Railroad Engineer Regiment 3, in Russia in October 1941.

Ford: An American Ford V-8 Model 78 (1937) with elegant whitewall tires, with the II./JG 54 in France.

Ford: The car of a commanding general of the Luftwaffe, a British Ford V8, 30 HP, Model 78 (1937), with masked headlights and right-hand drive. The British Ford V8 was practically identical to the American model.

Ford: This American Ford V-8 Model 78 (1937) made it to Smolensk in the winter of 1942-43.

Ford: How did the 216th I.D. manage to get this sleek American Ford V-8 "De Luxe Club Convertible Coupe" of 1938?

Ford: "Tunis, 1/28/1943, had good luck in British fighter-bomber attack", wrote Gerhard Kesenhagen (Panzer Grenadier Regiment 69 of the 10th Panzer Division) on this picture, which shows an American Ford V-8 Model A81 (1938) "De Luxe" (8 cylinders, 3.6 liters, 85 HP motor). The three dents in the roof and doors were marked with ink.

Ford: Lincoln was the prestige marque of Ford Motors, its standards of quality like those of the Cadillac. The picture shows a Lincoln Zephyr Sedan of 1938 with V-12, 4.8-liter, 110 hP motor in Oslo in 1941.

Ford: Under contract from the Netherlands Army, the DAF firm equipped a whole series of standard American Ford V-8 1-ton chassis with four-wheel drive and personnel-carrier bodies, mainly to tow light antitank guns. The picture shows a Ford 91 Y/DAF (1939) with the 1st Mountain Division in southern Russia in 1942, followed by a 1940 Chevrolet truck.

Oldsmobile: The 20th I.D. could enjoy life in France for only a few months. This is the car of the commander of Engineer Battalion 20, with the lettering "St" for staff and the pennant of an "independent battalion." The car is an Oldsmobile Type 60 of 1939, with a 6-cylinder, 3.5-liter, 90 HP motor. Oldsmobile is a branch of General Motors.

Packard: Dunkerque was a treasure-house of captured material that was serviced by the Todt Organization and made usable for the Wehrmacht. Here is an abandoned 1938 Packard, which will surely find a friend soon. Packard was one of the few independent automobile firms and – in constant competition with Cadillac – built only luxury-class cars with 6-, 8- and 12-cylinder motors.

Studebaker, another independent company, made military history chiefly with its legendary 2.5-ton trucks. The picture shows a 1937 Studebaker without any markings but a chalked "WH" on the right fender.

Studebaker Champion 6 of 1939, of the 216th I.D., seen on the "Grote Steenweg" (N 14) in Olesne, underway to the ferry across the Lys. In the background is a 1940 Chevrolet truck, followed by a 1938 Opel Olympia (or possibly a very similar Renault Juvaquatre." The precise locations come from Peter Taghorn, the author of "Mai 1940 – La Campagne des 18 jours."

Studebaker: Chief Designer Loewy created the typical "airplane" body at the end of the 1930s; this one is seen celebrating the 4000th mission flown by an unidentified reconnaissance unit. It is a Studebaker President 8 of 1939 with an 8-cylinder, 5-liter, 110 HP motor.

Willys-Overland: This firm is surely less well-known for its cars than for originating the legendary "Jeep." This picture shows WH-260 530, a Willys Custom Sedan of 1938 with 4-cylinder, 2.2-liter, 48 HP motor "at the waterfall on the road to Dray on 6/10/1942." The "shark nose" was the latest style on body design in its day!

Willys-Overland: 640,000 of the 1/4-ton four-wheel-drive vehicle with the (unofficial) name of "Jeep" were made by Willys and Ford by August 1945. The design and qualities of this "legend" cannot be dealt with in this book – countless books specialize in it. The picture shows a captured Jeep in Rome in the autumn of 1943, with high-ranking and highly decorated occupants: Field Marshal Albert Kesselring (Commander of the southern front), at left rear, Oberstleutnant Karl-Lothar Schulz (Commander, Paratroop Regiment 1), beside him, Generalleutnant Richard Heidrich (Commander, 1st Paratroop Division).

Willys-Overland: Among their many shipments to help the Red Army, the USA sent great numbers of Jeeps to Russia.

This Jeep was captured near Kharkov by the 2nd SS Panzer Division "Das Reich."

The car had a 4-cylinder, 2.2-liter, 54 HP motor which, like all American types, is said to have had high fuel consumption – for the sake of truth it must be said, though, that the comparable European vehicles of the time also had a powerful thirst!

Appendix

The Towing Axle

Necessity is the mother of invention – cars were turned into towing tractors:

On June 4, 1940, the Reich Traffic Ministry issued a directive concerning the modification of passenger vehicles into makeshift towing tractors:

I. To alleviate transport difficulties, large quantities of passenger vehicles will be modified into towing tractors in the near future by makeshift measures, in that an auxiliary axle equipped with a special frame and powered by chain drive offset from the rear axle of the vehicle, will be installed under the vehicle. Passenger vehicles modified in this manner are to be regarded as towing tractors. The original use permit issued for the vehicle will be voided. A mew use permit must be applied for. In the vehicle's papers as well as all documents and reports, these vehicles are to be designated at towing vehicles (tractors).

II. . . . Top speed 20 kph . . .

III. The auxiliary axles require permits!

Fritz Wittekind wrote about them in "Motor und Sport" in July 1940:
"The Beco axle of Chief Engineer Heidemann is intended first of all for the appropriate modification of smaller passenger cars. The first series were made originally for the Ford Eifel. Others are being prepared for certain types of Opel and Hanomag. In this way, such cars can tow a total towed weight of approximately 5.5 tons. For the installation of the Beco axle in passenger vehicles, no changes of construction are necessary, and the fenders do not need to be removed. It is sufficient to shorten the rear fender somewhat at the back after removing its normal rear wheels. In cars with thermosyphon cooling, an additional water pump is installed behind the radiator. Then it is also necessary to install an RPM regulator which limits the top speed of the motor to the speed at which the motor produces its most favorable torque. For example, in the Ford Eifel this is the case at approximately 2500 RPM. The complete modification can be made in less than half an hour. Finally, it must be stressed that this kind of makeshift tractor is to enjoy complete freedom from taxation for the duration of the wartime economy, according to a special directive of the Reich finance ministry. The Beco axle is presently the only auxiliary towing axle that is available for such modifications. To be sure, other designs of this type are also in progress. But this work has not yet been concluded to date."

The picture shows a Ford Eifel (1937-1939 model) with auxiliary towing axle and gas tanks to provide fuel.

Road Signs

The Wehrmacht's typical road signs were a big help to the driver –as well as for partisans, who were thus informed of the locations of large units, ammunition supplies, etc. The "Axle Street" must have been in the Rostov area.

Baue Deinen Wagen hier an dieser Stelle auf, damit sich jeder des Nachts den Schädel einrennt. Es ist ja Krieg.

Hier sank samt Karren in den Grund
Ein Fahrer –
Er fuhr Spur, Der Hund!

Auf 40 Meter Abstand achtet! Ihr habt die Rollbahn nicht gepachtet!